ROGUES AND SCHOLARS

JAMES STOURTON is a British art historian, a former Chairman of Sotheby's UK and the author of *Great Houses of London*, *British Embassies* and the authorized biography of Kenneth Clark. Stourton has lectured to the Cambridge University History of Art Faculty, Sotheby's Institute of Education and The Art Fund, and is a senior fellow of the Institute of Historical Research. He sat for a decade on the Heritage Memorial Fund Panel and the Acceptance in Lieu Panel, the two main conduits for art passing from the private to the public sector.

Also by James Stourton

Kenneth Clark: Life, Art and Civilisation

British Embassies: Their Diplomatic and Architectural History

Great Houses of London

Heritage: A History of How We Conserve Our Past

ROGUES AND SCHOLARS

Boom and Bust in the London Art Market 1945–2000

JAMES STOURTON

An Apollo Book

First published in the UK in 2024 by Head of Zeus Ltd,
part of Bloomsbury Publishing Plc

9 7 5 3 1 2 4 6 8

A catalogue record for this book is available from
the British Library.

ISBN (HB): 9781804541975
ISBN (E): 9781804541951

Typeset by Divaddict Publishing Solutions Ltd

Printed and bound in Great Britain by
CPI Group (UK) Ltd, Croydon CRO 4YY

Head of Zeus Ltd
5–8 Hardwick Street
London ECIR 4RG

WWW.HEADOFZEUS.COM

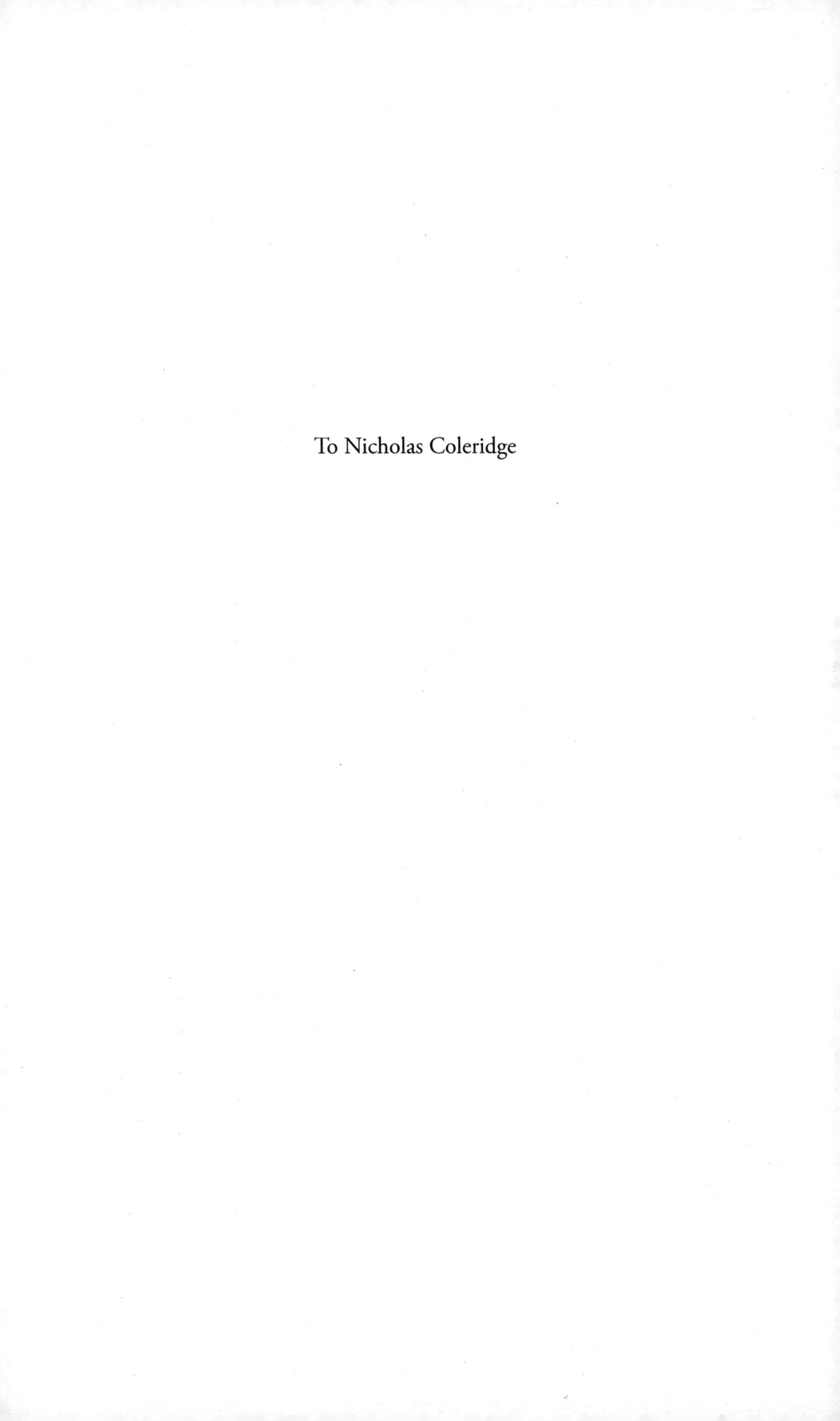

To Nicholas Coleridge

Contents

Preface

In 1966, John Berger, author of the highly influential art book *Ways of Seeing* (1972), wrote in the *Times Literary Supplement*: 'The story of how London has become one of "the world's capitals of art" would be fascinating – if it could be told. But probably it never will be told because there are too many well-kept secrets. One might as well try to tell the story of the last ten years of the Casino at Monte Carlo.'[1] I have rashly attempted that task.

I worked at Sotheby's for thirty-two years, mostly in London, initially in the British Paintings Department and latterly as the UK chairman. When I turned up for my first day at work in June 1979, Sotheby's was in the afterglow of the Mentmore and von Hirsch sales. Peter Wilson, who hired me, seemed all-powerful, and London unassailable as the world capital of art auctions. This book describes how that came about and why it changed over time.

My focus is on the auctioneers and dealers who contributed to London's global dominance in the second half of the twentieth century. I have not written about rare books, manuscripts or the jewellery trade (excepting S. J. Phillips in the chapter on silver), all of which also supported London's ascent. The difficulty in writing a book of this sort is to avoid becoming a gazetteer, and therefore many excellent dealers have been left out. I hope they will forgive my omission – perhaps they will be grateful.

The literature on the London art trade 1945–2000 is thin. There are several books about Sotheby's and some good memoirs from Christie's directors: John Herbert, Christopher Wood (unpublished), Brian Sewell and Charles Hindlip, not to mention the glorious writings of Philip Hook, who experienced the trade from several perspectives. When writing about the dealers of the era, I have relied heavily on oral histories from those I interviewed, all of whom have been generous with their time and knowledge. What shone through was their enthusiasm to discuss the past and remember the brilliance, conviviality, passion and occasional peccadillos of their colleagues over the last half-century. Many will agree with Lord Hindlip, Christie's late ex-chairman, that 'I doubt they have as much fun today as we did then'. Times change, however, and readers of the YBAs chapter will see that they had just as much fun at the turn of the millennium – but it is different.

Introduction

'A good art dealer is a man who can sell a picture he dislikes to a client who dislikes it'.[1]

Arthur Tooth in conversation with Henry Roland

The modern art market was born in London at 9.30 p.m. on 15 October 1958. That night, Sotheby's staged a glittering sale of impressionist paintings imported from America from the collection of the late Jakob Goldschmidt. Masterminded by Sotheby's newly appointed chairman, Peter Wilson, the sale introduced many novel features at a time when auctions were still trade affairs, with the air of the old curiosity shop. Held in the evening and in black tie, there were movie stars in the front row, television cameras rolled and, most startling of all, there were only seven items for sale. Goldschmidt was the first 'event' sale of the modern era and it was clear that a great change had taken place: Wilson had seen the future and grasped the importance of public relations. The sale established London as the global centre of the art market and Sotheby's as an international auction house. It began a shift in power from the dealers to the auctioneers and signalled the dominance of impressionist and post-impressionist paintings over the art market for the next forty years.

London after the war was like an Aladdin's cave, its antique shops stuffed with a quality and quantity of goods almost inconceivable today. These treasures came from the sale of country houses and the unloading of 200 years of obsessive collecting by a once rich country that found itself much poorer after the war and in need of cash. London had the goods when it came to old masters and furniture, but the art that would give the city global domination had to be imported, often from America. It was this coup that Peter Wilson pulled off in 1958 with Goldschmidt, enabling London to capture the impressionist market from Paris and New York.

More than anybody, Peter Wilson set the compass for the times. Geraldine Norman, the veteran saleroom correspondent, described his impact on the art market as mirroring the innovation 'of Henry Ford on the mass production of cars, Rothschild on banking or Fleming on medicine'.[2] Wilson's idea of persuading Americans to sell French paintings in London (where there was no indigenous market for them) back to Americans was enabled by technology, particularly the jet aeroplane and the telephone, and assisted by the advantages of geography, a common language and GMT. He saw an opportunity and seized it. Today, largely unknown outside the older members of the trade, Wilson is absent from the American-focused Sotheby's Wikipedia entry.

The 400-year history of the art trade is characterised by the tug of war for control of the market by dealers or auctioneers. Dealers thrived when cash was king, and controlled the Grand Tour, the French Revolutionary period, the Napoleonic wars and the interwar era, dominated by Lord Duveen's hold over the market. In the forty years after the Goldschmidt sale, the art trade became an auctioneer's market, until the rise of contemporary art around the millennium returned some of the power to the dealers. As a critic noted in 1965: 'The fact that you sold your flower painting in the saleroom is significant. Ten years ago, you would almost certainly have taken it to a dealer.'[3] By 1969, it was estimated that the turnover of London salerooms exceeded those of New York and Paris combined.[4]

Before the Goldschmidt sale, the international art market was operated by a cadre of major art dealers who kept offices in Paris, New York and London. The pre-war dealers' practice was to buy from private collections and sometimes replenish their stock from auction. In the 1950s, neither Sotheby's nor Christie's had yet expanded beyond London. During the period of this book, their story emerges as one of the great business duopolies. Likened to Tweedledum and Tweedledee by the dealer Lillian Browse, Sotheby's and Christie's became as aggressive, dominant and competitive in their field as Pepsi and Coca-Cola. Increasingly, collectors would offer their collections to the auction houses for sale rather than the dealers.

The difference between Sotheby's and Christie's was often perceived as one of social background, and although there was an element of truth in this, it was more a contrast in attitude that defined them. The departments at Sotheby's were less well-connected in society, making them more aggressive in scouring for business. Philip Hook recalls that, when he joined Christie's in 1973, 'One crusty old Christie's director explained to me that yes, Sotheby's sales were larger, but their approach was "dangerously commercial". In his eyes there was something ungentlemanly about making too much money.'[5] However, Christie's proved to be quite as ruthless as Sotheby's when the two firms imposed the buyer's premium in 1975. The introduction of the buyer's premium turned out to be a second caesura moment in our story. The rise of the two main auction houses to a position of dominance meant that firms like Agnew's had to reinvent themselves as the bidders of choice for the world's top museums.

If Sotheby's and Christie's are in some sense the lynchpins of this story, the art trade as a whole presented a glorious gallery of characters, comprising clever amateurs, scholarly specialists, brilliant émigrés, grandees with a flair for trade, knowledgeable stallholders, cockney traders and a fair sprinkling of rogues. Part of the fun of the London art world was its variety and extent. This wider network formed the base of the pyramid by which objects would be transferred through five or six hands, as they passed from country dealers to specialists in Bond Street, Kensington Church Street or Portobello Road. As a backdrop to this were all the ancillary aspects of the art trade: fairs, restorers, academics, museum curators, restaurants and gentlemen's clubs. More than anything else it was the street markets that captured the imagination of the public, writers, filmmakers and pop stars and drew them into the orbit of the trade.

David Sylvester suggested that:

> Art dealers are promising subjects for biographies. They buy and sell portable objects that can easily cost more than a castle or two. They survive by outwitting some of the world's most cunning and ruthless manipulators of wealth, and they also know how to charm the old rich, key sources of supply. When they deal in the work of living artists, they shape the careers of some of the most charismatic and paranoid individuals of their time. Their operations tend to sail dramatically close to the wind of commercial ethics and sometimes of the law. And, having lived lives generally presumed to have been even shadier than they were, they may well be immortalised as builders of temples of high culture.[6]

This is as true today as it was in Lord Duveen's time.

In keeping with Sylvester's view, Sotheby's was particularly prone to scandal. As the historian Robert Lacey put it, the firm 'had no monopoly on amorality, but as in so many other areas, they practised it better than anyone else'.[7] Christie's by contrast appeared more careful. An example is the story from a member of staff about Christie's Argentinian representative illegally sending artefacts to New York for sale. Aggrieved because they were reported in the American financial results and he was not getting the credit, he complained to Jo Floyd, the London chairman, who, with his hands over his ears, repeated 'I have not heard this, I have not heard this!'[8] Illegal export, particularly of antiquities, is the theme of several chapters of this book and generated more scandal than any other area. It is a sad fact that some of the most talented figures described here are also the most problematic, their reputations tarnished by their involvement in trafficking looted objects and other offences.

The illegal auction 'rings' that were a problem during the 1960s appear in several chapters. A ring involves a group of dealers clubbing together to avoid bidding against each other at auction. One dealer would be designated by the ring to bid for the object, which would then be re-auctioned in a private 'knockout', often held in a local pub after the sale. If knockout rings were the bane of the auctioneer's life, 'taking bids from the chandelier', by pretending there was a bidder, was equally enraging to dealers and collectors in the room. For the auctioneer, this was a legal method of getting momentum up to the reserve, after which they were obliged to accept only genuine bids. That said, rare rogue auctioneers, sometimes aware of a written bid left with the sales clerk, would 'run up' the buyer. The trade had a shrewd intuition about such shenanigans and constructed ruses to catch them out.

Despite its occasional scandals, London was considered to have the best and most reliable expertise in the world, and the city nurtured scholarly dealers of great repute and honesty. The exposure of the fake Samuel Palmers in the Keating affair marked a turning point after which the trade recognised the necessity of relying on something more rigorous than overly optimistic connoisseurship. Specialist markets like porcelain and drawings thrived with superb expertise allied to high ethical standards, whilst the evolution of art-fair vetting committees was also important in raising standards all round.

If there was one factor which improved the expertise of the London trade it was the number of émigrés who established themselves in the city, bringing a cosmopolitan sophistication. For many observers, this group of post-war exiled dealers 'created the civilised atmosphere of the London art trade in the 1950s and 1960s'.[9] Hans Gronau, Herbert Bier, Erica Brausen, Frank Lloyd, Harry Fischer, Hans Calmann, Annely Juda, Henry Roland, Hans Backer, Andrew Ciechanowiecki, Jacques Koopman, Julius Weitzner, Ralph Nash, Helen Glatz, Herbert Rieser, Hanns Weinberg and many others appear throughout this book. Mostly (but not all) Jewish refugees, a remarkable number had been interned on the Isle of Man during the war and subsequently served in the Pioneer Corps. They gave London an edge, animating the market with their incomparable knowledge and energy.

For Sir Geoffrey Agnew, 'Art dealing is rather like operating an intelligence network ... you have to know before anyone else when important pictures are coming onto the market.'[10] Finding the items for sale is the prerequisite of great dealing. When Paul Mellon once asked his father, Andrew, why he dealt with such an obvious charlatan as Lord Duveen, the older man answered bleakly, 'He has the goods.' Circulating to meet clients, and picking up whispers, and information, made 'charm, power of personality, and cunning part of the armoury of the supreme art dealers'.[11] For Richard Herner, 'Dealers were like butlers to the rich', and holding on to a good client might make the difference between profit and insolvency. It is extraordinary how many dealers were dependent on one major collector for a time. The client's collecting span might be seven years or, in rare cases like Paul Mellon, continue for forty years. As Jim Lally, a New York dealer in Chinese art, exclaimed, 'If you have one client, you are doing well, two and you are a hero.'[12]

Galleries were prey to timewasters, nutcases and fraudsters, some simply wanting to say they had a better one at home, others pretending to be millionaires or dukes, or to be furnishing rooms, and then disappearing. They would invariably be greeted by an impeccably turned-out receptionist.[13] Until the 1970s, in shops like Mallett's, 'discussing prices was virtually unheard of. The required sum was on the ticket, as it is today, and that was it.'[14] It was something of a novelty, therefore, when one day Mallett's director, Lanto Synge, encountered an Arab client who made an offer on two pieces of furniture. He was about to leave the shop when he was invited back for a conversation. For Synge, 'That was the beginning of a new way of dealing and a new term, "special price" was conceived.'[15]

Museum personnel were divided in their attitude to the trade. There were those who enjoyed coming to sales and keeping abreast of what was going on, and those who maintained a *de haut en bas* attitude. When Philip Pouncey, the British Museum's old master drawings scholar, joined Sotheby's, many of his former colleagues complained about his move to the 'dark side'. The Courtauld Institute, the powerhouse of British art history, shared this view. Despite providing Brian Sewell for Christie's, Anthony Blunt, when director, exhorted his students to have nothing to do with the art trade. By contrast, the Cambridge History of Art faculty, set up by Michael Jaffé, was trade-friendly (some thought too much so) and from the 1960s onwards provided many of the senior figures at Sotheby's and Christie's. Up until that point, Peter Wilson had favoured taking on bright boys straight out of school, and as Rupert Maas points out, 'you never lost any money by not having an art degree in the art trade'.[16]

In the 1970s, the entry point for men in an auction house was usually to start on the front counter or as a porter and then work their way up. For women, it was invariably the secretarial route. At Sotheby's, which was ahead of Christie's in this regard, several departments were run by women by the early 1980s. However, it would not be until the formation of a graduate trainee scheme at the end of the decade that greater gender parity was achieved.

The route to becoming a dealer was manifold. Some belonged to the great art-dealing families, while many more started as traders in Portobello, especially in areas such as tribal art, porcelain and antiquities. Other dealers gained their training from the auction houses. One of them, Adrian Eeles, describes the shift from one camp to the other, recalling:

> I quickly found that the transfer from auctioneer to dealer was far less easy than I assumed. One surprise was to find how reluctant some collectors were to commit to a purchase, very often haggling and then walking away without any decision. I was used to bidding in the sale room (often for the very same people) being more decisive. Museum curators were usually more focused, but their follow-ups often petered out after too long a wait.[17]

In the Christie's porcelain department, Anthony du Boulay was fond of saying that 'Fashion is the most important thing in the art market', and the period is marked by several revolutions in taste. London dealers were at the forefront of the baroque revival in paintings and sculpture, as well as the interest in Victoriana. Undoubtedly, the headline story is impressionism's extraordinary rise in popularity from the 1950s onwards, reaching its zenith in the market during the 1980s, when post-impressionist works overtook them. Perhaps the most astonishing change to the market emerged during the 1990s with the rise of contemporary art. At a time when major post-impressionist sales had decamped to New York, and old masters were drying up, the Young British Artists (YBAs) helped to maintain London's international relevance in the trade.

Various trade associations operated in the unregulated market. The most important were the British Antique Dealers Association (BADA) and the Society of London Art Dealers (SLAD). BADA dominated in the 1950s and 1960s, negotiating the best terms for trade and export licences with the government. Jeremy Cooper parodied BADA as 'pen pals with the editor of *The Times*: "Save Our National Heritage" they cry when arguing for the sale of their most expensive items to a British museum; "Export Antiques and Earn Dollars" they cry if asking for a licence to export the same piece to an American museum.'[18] Without question, BADA's lowest moment was its members' involvement with the Northwick Park ring scandal in 1964. Its greatest achievements were the organisation of the annual Grosvenor House Fair and the creation of its vetting committee. SLAD was set up to help and protect members, while instituting a buyers' dispute mechanism for referrals.

Perhaps the trade bodies' most useful government intervention was over the introduction of VAT, enabling them to maintain the margin scheme on profit, rather than full value, from 1986 onwards.* This was a huge gain, although the arrival of 5 per cent import VAT in the late 1990s was a blow to London as a trading centre. This tax was, however, deferred on arrival from outside the EU to the point of sale, and only applied if the object came into the EU. Europe's introduction of *Droit de Suite*, or the Artists' Resale Right (giving artists or their estates a small fee on the resale of art that continued for seventy years after their deaths), also appeared in the 1990s. It ultimately worked to the benefit of the American market, by making London a less attractive marketplace.

* LAPADA – the London Association of Art & Antiques Dealers – was founded in 1974 to oppose the VAT special scheme.

Throughout the period, the boom-and-bust cycle of the art market – usually five to seven years up followed by two years down – is set against the wider economic shifts. It is a truism that the art market continues its ascent after a financial market collapse, serving as a last 'refuge' for capital. After the 1929 crash, this was dubbed the 'tarantula theory' as a spasm of activity before sudden death. It would usually take a year for such economic shocks to filter through to the art market. After a momentous crash in 1974, Christie's laid off twenty-six staff and directors took a voluntary 10 per cent salary cut. Sotheby's made thirty-one redundancies. The immediate result was the imposition of the buyer's premium. By the Mentmore and von Hirsch sales in 1977 and 1978, the market was back in rude health. True to the cycle, another dramatic crash came in 1990, when the Japanese left the market, which coincided with the first Gulf War.

In the end, it is by the goods and the expertise that a market is judged, and what a procession of masterpieces London supplied during the period between 1945 and 2000. Starting with the Goldschmidt sale, London sold some of the finest impressionist paintings, including Cézanne's *Garçon au Gilet Rouge*, Monet's *La Terrasse à Sainte-Adresse*, great old masters, including the Velázquez *Portrait of Juan de Pareja*, Titian's *Death of Actaeon*, Mantegna's *Adoration of the Magi* and Rembrandt's *Titus*, the contents of Mentmore, the treasures of Robert von Hirsch, the drawings from Chatsworth, Van Gogh's *Sunflowers*, Adriaen de Vries's *Dancing Faun*, Picasso's *Weeping Woman*, and Damien Hirst's shark. You have only to look at the formation of the Getty Museum to comprehend the impossibility of its existence without the supply of the London art market. As Sherman Lee said at the fiftieth-anniversary celebration of the Cleveland Museum, 'When a curator comes back from a trip and says, "Look what I have found!", I say to him, "Remember, a dealer found it before you."'[19]

Chapter 1

The Goldschmidt Sale

'The detonator of the huge price explosion was the sale of Impressionists from the Goldschmidt collection ... from that moment, the art market changed and ethically not for the better.'[1]

Lillian Browse

Up until World War II, Christie's was by any measure the greatest art auction house. With a long and distinguished history stretching back to James Christie, who founded the firm in 1766, Christie's had an untouchable record of selling great masterpieces from Botticelli to Rembrandt. Whilst historically they had occasional rivals, notably Phillips in the later eighteenth century, Christie's saw it as their destiny to serve the British aristocracy in a cosy monopoly, with Sotheby's dealing with the contents of the library. Sotheby's began to challenge this arrangement between the wars when they expanded their operations into works of art and paintings. So successful was the challenge that, in the decade after 1945, Sotheby's and Christie's were already evenly matched in turnover, although Christie's had a slight advantage in six out of the ten years. It was not until 1956 that Sotheby's pulled ahead dramatically (Christie's £1.68 million vs Sotheby's £2.27 million) and the gap would widen further to give Sotheby's the dominant market position until the 1980s.

After being destroyed during the Blitz, Christie's premises were beautifully rebuilt with the new octagonal saleroom more adaptable to modern conditions. Given the later fierceness of the duopoly, it is a surprise to discover that in 1947 their partners approached Sotheby's with a view to amalgamation, the third merger proposal in fifteen years. Negotiations did not proceed; Sotheby's partners were already sensing an ambitious future, while Christie's chairman, Sir Alec Martin, saw no cause for alarm.

Given Christie's grand reputation, Alec Martin's background was remarkably humble. The eighteenth out of twenty children of an East End shopkeeper, he came to Christie's in 1897 at the age of twelve, as a good-looking office boy whose principal responsibility was to fetch pints of beer for the head porter.[2] Most of the porters had been 'gentlemen's gentlemen', but Alec had ambition and took himself to Paris, where he washed dishes to pay his way. He would arrive at Christie's early to view the sales and caught the eye of two art world grandees, Claude Phillips, the first keeper of the Wallace Collection, and the dealer and impressionist collector Hugh Lane. Surprisingly, given the friendship with Lane, Alec Martin never saw the point of the impressionists and at first would transfer them to his friend, the dealer Oliver Brown, at the Leicester Galleries.* This indifference would cost Christie's dear. Anthony Lousada, the firm's solicitor, remembers walking round the rooms viewing Picasso and Braque with Alec Martin, who referred to them as 'This 'ere filth'.[3]

Martin also gravely underestimated his rivals in Bond Street. The Christie's silver expert Arthur Grimwade believed that 'Alec treated Sotheby's as a fly to be brushed aside … and we were all inculcated with the same attitude.'[4] To Brian Sewell, then a

* Oliver Brown is one of the great pre-war dealers who heroically promoted modern art in London from his modest gallery.

young specialist in the Christie's Old Master Department, who tended to hyperbole, Alec Martin was 'a vile, arrogant and ignorant old martinet who had obscured his working-class origins and converted his Cockney speech into a post-Edwardian affectation'.[5] Sewell was no doubt offended by Martin's dictum: 'remember, we are not experts, we are auctioneers'.[6] However, Martin achieved what no other auctioneer has ever attained, a knighthood, and the address at his memorial service was given by Edward Heath, the then Prime Minister.

In the post-war years, Sotheby's had about 120 employees and its focus was still on book sales. Their chairman, Charles des Graz, was learned and knowledgeable, with a strong interest in heraldry. The partners realised that to expand they needed to pursue old master paintings, the big game of the art world, more vigorously.* One of the partners ran into Sir Kenneth Clark in the street and asked his advice. Clark recommended Hans Gronau, a scholarly German art historian who had just been demobbed from the Pioneer Corps (whence came so many in this story). Gronau had married a fellow art historian, Carmen, to whom he had been introduced by Nikolaus Pevsner. He had a weak heart and would turn up at Sotheby's accompanied by his wife, who gradually took over his responsibilities. Immediately recognising her quality, Peter Wilson promoted her cause. He and Carmen Gronau certainly looked an odd pair: she short and direct with an accent, he tall, cherubic and soothing. They were to remain the closest of corporate allies until Wilson's retirement in 1980.†

* Between the wars, Sotheby's had hired Tancred Borenius, the distinguished Finnish art historian, as a consultant in old master paintings.

† Sotheby's had already had a female partner in the 1930s; this was Evelyn Barlow, who took auctions and had wide responsibilities.

Wilson's family were Yorkshire landowners. His father, 'Scatters' Wilson, was a notorious rake and gambler, his mother, Barbara Lister, the cultivated Francophile daughter of Lord Ribblesdale, famous from his commanding portrait by John Singer Sargent. Peter was an amalgam of his parents: he inherited his father's 'chancer' side and his mother's refinement and cosmopolitan interests. He had a conventional upper-class education, Eton and Oxford, but was a late developer and didn't shine at either. His first taste of the art world was with the dealer Spink, which he found 'cold, snobby and beastly in atmosphere'.[7] A period at Reuters followed and a spell selling advertising space for *Connoisseur* magazine. Wilson was hired by Sotheby's in 1936 as a trainee in the furniture department and two years later became a partner in the firm, buying 5,000 shares with his wife Helen's money. She was a German woman of bourgeois origins who gave him two sons. The marriage didn't last – Peter was gay – but they remained on good terms and he was a devoted father.

Wilson had spent much of the war in America working in postal censorship. Once he became well known, the notion that he had been a spy made an irresistible story. The mischievous suggestion was put about that he might be the 'fifth man'. When Wilson was questioned about this he said 'Really? Just stop and think, I am completely conservative. I am interested in people who are rich and noble. And do you think I could be working for the Soviet Union? It doesn't make sense.'[8] Naturally, he knew Guy Burgess, as everybody else did, and was a friend of the art-dealing spy Tomás Harris.

Returning to Sotheby's at the end of the war, Wilson showed a natural leaning towards objects. He produced some scholarly catalogues, notably that of the Guilhou antique rings, quoting a wide range of literary sources from Pliny to Jean-Jacques Rousseau.[9] There is no doubt that his personal interest remained with objects and sculpture, but given the trajectory of the company, Wilson was made head of the paintings department. This meant pursuing old master paintings in the teeth of Christie's superiority in this field. His mettle was to be tested in 1956, when Sotheby's was offered Poussin's *The Adoration of the Shepherds*.

Great old master paintings still tended to be sold privately by dealers and the Poussin was something of a test case. The painting's owner, Commander Beauchamp, was a Norfolk landowner who claimed to have been offered £10,000 cash from a dealer, a figure which soon turned into £15,000. Wilson offered Beauchamp a guarantee of £15,000, meaning that whatever the outcome of the auction, the seller was guaranteed to receive this as a minimum sale price for the work. It was the first such offer in Sotheby's history and a considerable risk.

Beauchamp's butler duly delivered the picture to London. After a visit to his club, however, Beauchamp raised his price, asking for a reserve (and guarantee) of £35,000 or he would withdraw the painting. Sotheby's partners, who didn't like the original guarantee, were now very anxious, but Wilson was determined to push ahead, telling colleagues, 'If we fail over this, we are done for.'[10] Anthony Hobson, the most conservative of the partners, objected: 'It seemed to me not altogether how auctioneers ought to behave.'[11] It was certainly the beginning of a very slippery slope and a foretaste of important auctions to come. At the sale, Wilson was the auctioneer and announced lot 119 as 'The Poussin'. The bidding started at £1,000 and the picture sold for £29,000, a loss of £6,000 for the firm, a considerable sum in those days. Bought by the London-based dealer David Koetser and underbid by Agnew's, the painting fetched the highest price of the season. The case demonstrates Wilson's ruthless determination to break Sotheby's into the old masters market, whatever the outcome. Whilst he achieved that, it may also have persuaded him to look elsewhere for growth.

Wilson's internationalist outlook made him interested in the burgeoning area of impressionist paintings, to be found mostly in America and on the continent of Europe. Sotheby's had separated old master sales from impressionism in 1951, something that would not happen at Christie's for another thirteen years. Relishing the company of collectors, however eccentric, Wilson began to travel extensively to gain the consignments in this area, which he saw as the key to the future. An excellent linguist, fluent in French and German, he preferred dealing with knowledgeable owners. He once remarked:

> I would much rather deal with a Jew than a duke. If I tell a Jewish collector that I think I can get thousands more for a picture than he expected, he sees me as a partner. He'll cut me in on the deal. But if I tell a duke he owns a masterpiece that he never suspected, his first thought is whether he can cut down my commission on the extra money.[12]

The two years following the Poussin sale were decisive in establishing London as the place to achieve the highest prices for impressionist paintings. At first it looked as if Paris, where several fabulously rich Greek ship-owners had settled, might be the hub. The most famous among them – Goulandris, Embiricos and Niarchos – had made the first great European fortunes after World War II by buying up redundant Liberty ships and cornering the world's shipping market. After the closure of the Suez Canal in 1956, their profits escalated. In May the following year, Basil Goulandris paid almost $300,000 at the Galerie Charpentier in Paris for Gauguin's *Still Life with Apples*. The Greeks offered stiff competition to American collectors, which lent weight to the argument for selling in Europe. Most rich Americans came to Europe to shop every summer but Europeans were less likely to go to New York, so the argument ran. Although Paris held one more great impressionist sale, the Cognacq Jay collection, the French market was held back by the

Commissaires-Priseur system that governed auctions. Designed to protect the internal market, it became a fortress in which it was imprisoned. At this point, Sotheby's produced the double punch that gave London the key advantage.

The British government's lifting of the strict control on the export of sterling after 1953 made it easier for London to pursue international business. Both Sotheby's and Christie's began scouting trips to America, their selling point being London's lower commission rates – 10 per cent as opposed to 20 per cent at the grand old New York auction house, Parke-Bernet. Sotheby's interest in America paid off when Wilson and Carmen Gronau were commissioned to sell fifty-six impressionist paintings from the Weinberg Collection. The auction, held in London in July 1957, established Sotheby's credentials to hold the Goldschmidt sale. Weinberg Jr had cannily put in the contract that Sotheby's must retain a PR firm. They hired the American advertising agency, J. Walter Thompson, who put 'The Queen' at the top of their party invitation list (which most assumed meant the magazine). When the monarch received the invitation, she turned to her private secretary: 'We've never been to one of these before. Why shouldn't we go?'[13] Noting that Queen Elizabeth, Prince Philip and Princess Margaret attended the private view, a *New Yorker* headline ran: 'Sell at Sotheby's and get to meet The Royal family'. The sale made £326,520, Sotheby's highest total to date.

The Weinberg sale achieved international attention, not just in the press but on radio and television. The BBC's *Panorama* dedicated a whole programme to the event, and the American *Christian Science Monitor* announced that 'London has become the centre of the art market'.[14] For Frank Herrmann, 'it established the brilliance of Peter Wilson, the auctioneer as impresario … he was able to see ahead with a clarity and a degree of assurance an immeasurable advantage over its rivals'.[15] It was in 1958 that everything changed: this was the year that Wilson became chairman, the year of the Goldschmidt sale and the year that the direct jet aeroplane service started between London and New York. Two essentials of Wilson's success were the jet and the telephone, and he was rarely off either.

Although the Goldschmidt sale was Wilson's finest hour, Sotheby's started from a backward position. The banker Jakob Goldschmidt died in New York in 1955. It was left to his son Erwin, a difficult customer who enjoyed being awkward, to decide whether to sell his father's paintings in Europe or America. Following the success of the Cognacq Jay sale in Paris, he opened negotiations initially with Maurice Rheims, but found that expenses would have amounted to 35 per cent of the proceeds.* He and his co-executor, Jesse D. Wolff, therefore came to London and arranged to see representatives of both Sotheby's and Christie's in separate suites at the Savoy Hotel. Christie's had cleared the pictures through customs, so they may have been complacent. Understanding Erwin and Wolff's German, Wilson and Carmen picked up that Sir Alec Martin was in the next room. How relieved they must have felt when Erwin turned to Wolff, saying: '*Die hier gefallen mir viel besser*' – 'I like this lot much better'. As John Herbert, later head of Christie's Press Office, commented, 'this pronouncement was probably the most fateful to be made in the turbulent post-war history of the London art market'.[16]

Sotheby's strategy for the sale was simple: to express their confidence by offering very low commission up to the reserve price, with a rising scale of commission thereafter. One of the most radical aspects of the sale was having only seven lots, an act of veneration that was the very opposite of the wholesale tradition of auctioneering. The evidence is contradictory as to whose idea this was, but it was probably Wilson's, and no doubt

* Goldschmidt even entertained a bid from Stavros Niarchos (believed to be £600,000) for the whole collection.

he suggested holding it at 9.30 p.m., making it the first evening sale in London since the eighteenth century.[17] Attendees were asked to wear evening dress, unheard of for a London auction. There were celebrities present: Margot Fonteyn, Somerset Maugham, Kirk Douglas and Lady Churchill. Ringing up his father's friend, Lord Beaverbrook, Wilson explained that the sale would be the making of the London art market, and in this he was percipient. Unusually, he was five minutes late starting the sale, because Erwin had insisted on someone being ejected for making an offensive remark. The next day, the *Daily Express* reported, 'Tall, dinner-jacketed Mr. Peter Wilson, Chairman of Sotheby's and auctioneer of the night, climbed the steps of the pulpit-like rostrum in the green-walled main saleroom. Chubby-cheeked Wilson blinked in the glare of the massed TV lamps, ran his eye over the mink-and-diamond-dappled audience, rather like a nervous preacher facing his first congregation, and rapped firmly with his ivory gavel.'[18]

The sale got off to a good start, with the Manet *Self-Portrait* fetching £65,000 against the reserve of £45,000, and so it went on – the Van Gogh started at £20,000 and rose to £132,000 (against a reserve of £118,000). The pinnacle was lot 6, Cézanne's *Garçon au Gilet Rouge*. Wilson started the bidding at his reserve of £125,000 and it quickly became a contest between two New York dealers. The price had swiftly risen to £220,000 when Wilson famously asked, 'What, will no one offer any more?' causing a ripple of laughter. Reaching the highest price ever achieved by a picture at auction at that time, the Cézanne was bought by the dealer Georges Keller on behalf of Paul Mellon. The entire sale had taken twenty-one minutes and every record had been broken. The only work bought by an Englishman was Renoir's *La Pensée*, which was acquired by the dealer Edward Speelman on behalf of a property tycoon, Jack Cotton. The sale totalled £781,000, and at the end the audience stood up, cheered and clapped as if it was the climax of an opera.

The grand London dealers at once realised that a new power had been unleashed. To counteract this would be challenging. However, one dealer had an immediate benefit. While Wilson was planning the Goldschmidt sale, he had discussed it with the furniture dealer John Partridge, whose elegant premises were opposite Sotheby's on Bond Street: 'I don't know what I'm going to do. I have just got Goldschmidt to agree to have a seven-lot sale and I thought for the first time we would ask everyone to put on dinner jackets and dress up. I can stretch the sale out for an hour but what on earth do I do after that?'[19] The answer turned out to be a shared reception in John Partridge's gallery, as a result of which the dealer sold a collection of furniture.

The Goldschmidt sale even exceeded Weinberg in its press coverage. It was suggested that the radio programme *Stock Market Report* should round off prices with news from the art market. Wilson had no qualms about this idea and was wont to say: 'Art for art's sake is really awful rot.'[20] He would later initiate his own art market report with the *Times-Sotheby's Index* in 1967. 'The thing about the Goldschmidt sale,' as Wilson told one journalist:

> was that before it no work of art, no picture, had ever been sold at auction for more than £35,000 [the Poussin]. What it did of course was in the first place to transfer attention firmly on the auction front from New York to London. It established that it was possible for people to make up their minds on the spur of the moment to pay these vast sums for a picture or not. Until then it was always believed by dealers, and by a lot of other people who went in competition with us, that people would not do that – they would have to have time to think about the money involved and so on.[21]

For Christie's, it seemed like a bitter blow. Alec Martin's son, Bill, was appalled that even after the Weinberg sale his father did not understand the importance of impressionism, and as he exclaimed: 'We're sunk.'[22] He was right, for that year Sotheby's pulled away from their rivals with a turnover of £3,077,526 as opposed to Christie's £1,687,373, advancing to £5.8 million in 1959. Sotheby's inexorable rise had begun but, although its turnover would remain nearly double that of Christie's until the 1980s, Christie's nevertheless followed in their slipstream. It was the power of the duopoly that led to the rise of London to a position of world domination.

Chapter 2

Sotheby's Goes for World Domination

'There is an extraordinary confidence, verging on arrogance, about the way Sotheby's straddles the art world.'[1]

Jeremy Cooper

When Peter Wilson was offered the chance to compete for the sale of Dr Roudinesco's collection of modern paintings in Paris, the day of the appointment was 21 May 1968, which happened to be at the height of the student riots or *les événements*. France had shut down with a general strike; there was no public transport of any kind, and all flights in and out of the country were cancelled. Undeterred, Wilson chartered a Piper Aztec owned by the Duke of Richmond and took off from Goodwood, before receiving the message that Orly and Le Bourget air traffic controllers were on strike and both airports were closed. As there were no aircraft in the sky, nobody in the tower and nobody to tell them not to, they landed unchallenged at the deserted Le Bourget airport, and arrived only fifteen minutes late for the meeting. The client was so surprised by his appearance that Wilson signed the deal within five minutes.[2] Nothing would stand in the way of Wilson winning a sale, but, as time would tell, this resolve was offset by a carelessness in the way the business was run and some cavalier decision-making.

Famous sales made headlines, but a note of populism – and even vulgarity – was to become part of the Sotheby's brand. In 1958, Wilson employed an unconventional public relations man, a very unmilitary Brigadier Stanley Clark. Clark's aim was simple: to make Sotheby's the most famous auction house in the world and the place where everyone would want to come to buy and sell. Wilson pointed out that the firm relied upon wealthy clients and Clark retorted with an important counter-argument: 'I am not trying to teach you your business Sir, but one should remember for every ten people with £100,000 to spend there are a hundred million with £100 to spend.'* And thus was born Sotheby's ambition, to paraphrase J. F. Kennedy, to go everywhere and sell everything to everybody. It was nothing short of world domination, and Souren Melikian of the *Herald Tribune* called it 'launching the Sotheby's rocket'.† At Christie's, a good-looking press officer called John Herbert was appointed. Adored by the staff, his main problem, according to one colleague, was that the senior directors shunned publicity and distrusted the newspapers.[3] As a result, Herbert was always on the receiving end of directorial ire. Evelyn Joll, chairman of Agnew's, would refer to him as 'Herbie the half-wit'.[4] Historians must be grateful to Herbert, however, as he wrote a useful account of Christie's during these years, *Inside Christie's* (1990), which the wits dubbed '*Outside Sotheby's*'.

As many famous brands have discovered, the fastest route to being a market leader is through a dynamic duopoly, and so Sotheby's world domination would only be possible with Christie's competition. At a time of national decline, the two

* In 1966, on a turnover of £21.5 million, Sotheby's sold 75 per cent of lots for less than £100.

† In the 1960–1 season (October to July), Sotheby's were at £8.4 million to Christie's £3.1 million.

auction houses shone internationally as a British success story during the 1960s. If Sotheby's usually led the way, Christie's followed – great sales poured into London, records fell like ninepins and the pair fought ferociously for business as they opened offices all over the world in a relentless push for growth. The traditional Bond Street dealers – who had lost their dominant position – were appalled but consoled themselves with the availability of great works of art for sale in London and by the arrival of so many new clients during the sales period.

Peter Wilson's opposite number at Christie's, Peter Chance, was also appointed in 1958.* Both Etonians, the two Peters had very different styles, which reflected their firms well. Wilson was a cosmopolitan to his fingertips. Peter Chance was a charming but more staid figure, at home on the grouse moors of Scotland. Chairman of the Georgian Group, his tastes reflected this: he collected eighteenth-century mezzotints and Chelsea gold anchor porcelain. If Chance worked in a grand office and would leave it mainly to take an auction or lunch at his club, Wilson disliked White's, which was the centre of Chance's world. Instead he had a poky space behind the main saleroom – 'less a chairman's office than the sacristy behind the altar'[5] – and spent his day dropping in on departments to see what had come in, or jetting off to visit clients.

To the Chinese porcelain dealer Giuseppe Eskenazi, Chance was 'a starchy and unbending personality'. He is recalled with affection by his former colleagues, but with a sigh that he stood little chance against the Mephistophelean Wilson. Brian Sewell, who heartily disliked many of his colleagues at Christie's, had a fondness for Chance: 'sturdily built, almost my height, he was a pink and silver man, a middle-aged *Rosenkavalier*, pink of complexion, silver of hair and often of suit, prinked, perfumed and immaculate, his bearing military, given to doffing his overcoat

* In 1958, Christie's was re-formed as Christie, Manson and Woods Limited.

as he climbed the stairs each morning'.[6] One observer described him as 'a procession in himself'. An ex-guards officer, he would speak in military terms of the 'junior echelons', and in fact in the 1960s Christie's rather resembled a Brigade of Guards regiment: hierarchical, exclusive and stylish with an undercurrent of mischief and humour. The front counter was populated with the godchildren of the chairman, often Etonians. With a 150-year-old advantage among aristocratic British clients, Christie's saw little reason to change its habits, and was supremely confident in the loyalty of its clientele. Sotheby's would challenge this loyalty but in the short term would have to look mainly overseas and to the trade for its business.

The two chairmen viewed each other through a baleful lens: 'Leave nothing to Chance' Wilson would intone, while Chance reserved for Wilson his favourite term of abuse, 'the man is a swine'. There was an old joke in the art world that when somebody died, Peter Chance went to the funeral and Peter Wilson went to the house. Whilst Sotheby's was powering ahead with a turnover running at roughly twice the size of Christie's for almost three decades, Chance would answer 'maybe, but Christie's is more profitable'.[7]

At a time when chairmen still held most of the executive power, both men were autocratic. Whilst Wilson was an impresario whose power rested on success, charisma and not a little cunning, Chance was more like the regimental colonel whose power rested on immutable bonds of respect for rank. When Christie's director Hugo Morley-Fletcher was bold enough to suggest at a board meeting that a matter might be referred to the staff, Chance turned to his successor, Jo Floyd, and asked: 'What's that fella think this is, a bloody commune?'[8] Nevertheless, Chance was certainly more collegiate than Wilson, who increasingly brooked no opposition.

Wilson's range of knowledge never ceased to amaze his colleagues. Julien Stock recalls Wilson passing through the drawings department and glancing at a pair of unattributed German Rococo drawings: 'Julien, those are by Baumgartner.' Although Stock had never heard of the artist, of course, Wilson was right.[9] Being socially grand and knowledgeable, Wilson could meet great collectors on equal terms, whether they were Stavros Niarchos or Paul Mellon. He had a good understanding and sympathy for American collectors, even the tycoon Norton Simon, the most difficult of them all. He was later to organise Simon's wedding to Jennifer Jones, which bizarrely took place on a boat on the English Channel.

When the directors of Christie's wondered how Wilson had gained such a lead, they generally pointed to the more talented board at Sotheby's, full of great experts like Anthony Hobson (books), Jim Kiddell (ceramics) and Tim Clarke (works of art). Key to Wilson's strategy was to create expert departments with first-class knowledge, ahead of anything seen before in auction houses. As he told a reporter: 'we changed the whole face of the art market in the 1950s by going over to specialised sales'.[10] The rapid expansion of the firm into relatively new areas like modern painting also meant that there was a cult of youth at Sotheby's. As Marcus Linell wrote, 'in the period 1955–65 a lot of energetic young people were recruited to the company. It was more difficult to attract seasoned art experts, so Wilson took them young.'[11] Linell himself was a bright grammar-school boy who was offered a job when he was sixteen but said 'I think I'd better get my o-levels first'. When he did join, he found 'it was terribly exciting. It was like being in some sort of troika with Peter Wilson whipping the horses on and you were always panting along behind thinking, my god what a wonderful life this is.'[12] The cult of youth was never more evident than in the most important department of all, Impressionist and Modern.

Having turned Sotheby's into the place for Impressionist and Modern, Wilson oversaw the department himself. Christie's was still enjoying their old master paintings heyday and didn't turn their guns in this direction until the mid-1960s. More serious was foreign competition. Both Sotheby's and Christie's charged a lower commission than their American and French equivalents. But as Wilson always pointed out, you didn't go to the doctor who was cheapest, you went to the one who got the best results. He was fond of engineering deals like the 1963 sale of René Fribourg's collection of French decorative arts imported from America, offering to take it commission-free unless the sale total exceeded $1.5 million. After this amount, he proposed 15 per cent commission up to $3 million, reverting to 11 per cent on anything beyond that. In the event, the sale totalled almost $3.5 million.

Wilson picked two very different characters to staff the all-important Impressionist Department, the future novelist Bruce Chatwin (hired at twenty-one) and Michel Strauss (who was twenty-five). Chatwin was typical of Wilson's young, quick-on-the-uptake charmers whose mutation from Marlborough *ingénu* to Wilson clone amused his colleagues, who found him bumptious. His first job at Sotheby's was numbering the porcelain and glass lots as a porter in the Works of Art Department. As a colleague noted: 'He was completely hopeless at the task and never did develop the ability to concentrate on the mundane.'[13]

Wilson saw potential in Chatwin and placed him in the Impressionist Department, where he and Michel Strauss were pitted against each other. Chatwin was, in Strauss's words, 'upset and furious when I arrived. Perhaps what riled him so much was the fact that I was the first person with a History of Art degree to join the company and walk straight into a job as a junior cataloguer.'[14] The rivalry between Chatwin and the older and more sophisticated Strauss came to a head over the Somerset Maugham collection. They agreed to catalogue the paintings together, but Michel arrived to find that Bruce had been in the office since 5.30 a.m. and had catalogued them all on his own. Chatwin left the firm aged twenty-six to pursue his successful career as a writer. He maintained a malevolent view of Wilson for the rest of his life, referring to his former mentor as 'The Beast', and writing disparagingly of him in his short stories. Strauss may not have had Chatwin's charm or charisma, but he knew pictures far better and was to remain at Sotheby's until 2000, the lynchpin of the department. Such was the success of bringing the impressionist market to London that Lillian Browse realised that 'the days gradually passed when it was necessary to go to Paris three or four times during the year in order to buy French pictures for stock'.[15]

If Wilson favoured ambitious, young, non-university trainees, he put seasoned consultants alongside them to ensure that catalogues were authoritative. Sometimes these were scholarly museum retirees such as Philip Pouncey, from the British Museum, who had formerly worked at the National Gallery, and Neil Maclaren from the latter. The dealers were not enthused by this development. During the 1960s, they still thought of auctions as wholesale trade affairs and regarded it as their prerogative to find 'sleepers', in the form of unrecognised or miscatalogued paintings. There was much talk of over-cataloguing by the auction houses. Many of Wilson's consultants were experienced dealers who saw no conflict of interest in their position, and included his close friends Jack and Putzel Hunt, or John Hewett, the enigmatic antiquities and tribal art dealer who 'looked like a poacher in his beard and scruffy cap'.[16] According to Wilson's son, Philip, 'John Hewett was Peter's best friend and he encouraged him to be interested in both tribal art and antiquity'.[17]

Spending weekends at his pretty Queen Anne house, Stone Green Hall, at Mersham in Kent, Wilson enjoyed seeing Hewett, who lived close by. Whilst Chance and the Christie's directors were entertained at pheasant shoots on the great estates, Wilson explored antique shops with Hewett and the Hunts in what was 'the nucleus of Peter Wilson's off-duty family'.[18] As far as the dealer Cyril Humphris was concerned, this group was a little cabal 'who knew something that we did not know'.[19]

Sotheby's and Christie's astonishing growth in the 1960s and 1970s was the result of establishing offices all over the world to feed the two principal international salerooms in London and (soon) New York. Their great colonisation began modestly in 1967 with sales at Gleneagles in Scotland, and representative offices followed in most of the capitals of Europe. Christie's were ahead of Sotheby's in taking a sale in Geneva in June 1968, the first by a British firm on mainland Europe. Geneva turned out to be a successful location, particularly for jewellery sales. Generally, the two houses had a different emphasis in their representation. Christie's ran their offices abroad with grandees, while Sotheby's had grandees too but focused more on general valuations and salerooms, with regular auctions in Amsterdam, Monte Carlo, Zurich, Florence and Frankfurt. Sotheby's established its most successful Continental saleroom in Monte Carlo (1975), which provided a foothold for the sale of French-sourced items while circumnavigating the legal blocks of selling in Paris. Although the Monte Carlo office seems obvious in retrospect, at the time it seemed like an odd location, as there was no hinterland of art dealers. It was, however, convenient for Wilson, who owned a house along the coast near Grasse. He opened Monte Carlo with a triumphant sale from the collections of Baron Guy de Rothschild and Baron Alexis de Redé.

The saleroom growth across Europe was mirrored in Britain, where Sotheby's, true to its populism, embarked on an ultimately unsuccessful policy of buying local salerooms in Torquay, Chester, Pulborough and Taunton to increase market share during the 1970s. The firm was, however, unable to compete with local auction houses, which were quicker on their feet and could do things more cheaply. Christie's preferred the less effusive route of using landed county representatives, generally not very hungry or knowledgeable, but with an ear to the ground.

Sotheby's caught the mood of the swinging 1960s with its ambition – fearless of vulgarity – and the youth of its staff, with department heads in their mid-twenties. People would speak of Wilson's Academy of Auctioneers, but some of these young men were more controllable than others, and the word 'ruthless' became attached to Sotheby's. Wilson's 'smootherboys' were in many ways natural dealers, and many left to be just that. The turnover of staff was troubling, and it was clear that a more rational approach to recruitment was required, rather than appointing the first person who turned up. Art history faculties were still in their infancy at British universities.

In 1969, Sotheby's set up an art course, which was devised to find and train talented young people. Initially, the participants were each paid £800 a year and sent on a three-week trip to Venice. The course was so successful that somebody asked the question why are we paying them? Why don't they pay us? – and thus the Sotheby's Works of Art Course, later the Institute of Art, was born. Run by a remarkable dynamo, Derek Shrub, formerly at the V&A, the emphasis was on the decorative arts, knowledge much needed at Sotheby's and hardest to find in the universities. It was when the young Caroline Kennedy went on the Sotheby's course that it became internationally famous and a profit centre. Whilst it did produce three European chairmen, Simon de Pury, Henry Wyndham and Mario Tavella, the quality of its graduates was distinctly uneven.

Without any policy as such, Christie's was recruiting higher-quality staff. Having had lacklustre expertise for so many years, employing people on the basis that 'Eton and the Guards gets you a job at Christie's',[20] they began to recruit graduates especially from the newly formed Cambridge History of Art faculty. A clever young generation of graduates at Christie's was to close the gap with Sotheby's: David Bathurst, Brian Sewell, Hugo Morley-Fletcher, Noël Annesley, Hugh Roberts, Francis Russell, John Lumley, Christopher Wood and Philip Hook. Being mostly Oxbridge, they were naturally collegiate and stayers, which gave Christie's greater continuity in its expert departments. The only female Christie's director at this stage was Hermione Waterfield, who was not a graduate and came up the classic route for women from being departmental secretary. Although Sotheby's was ahead in female promotions, there was not much real progress in gender equality until the early 1980s.

One of Wilson's great gifts was as an auctioneer, and he was described as 'a great dealer negotiating from the rostrum', with *Vogue* calling his style 'sublime'. With a half smile always on his lips, Wilson pioneered a laconic, smooth British approach that contrasted with the more energetic American manner. One colleague described how, 'There was something feline about the quiet way he conducted auctions from the rostrum, so affably confident that the bidders seemed to compete as much for his approval as for the lot he was selling. Only when the hammer fell might they realise how smoothly he had been changing gear to ratchet up the price against them.'[21] At Christie's, Bevis Hillier described Peter Chance as a 'figure of monolithic dignity at the rostrum'.[22] In the 1970s, Patrick Lindsay became the main auctioneer, 'noted for his sartorial elegance and for the intimidating way he raises his right eyebrow to prise another bid from a client'.[23] Every bit as suave as Wilson, Lindsay would famously ask 'Another one?'

In the late 1950s, Wilson asked John Fowler of Colefax and Fowler to redesign Sotheby's main galleries. After crimson was rejected, he came up with a smoky shade of green that turned out to be a good background for works of art and was associated with Sotheby's for the next thirty years. Aspiring to the Tribuna in Florence, Christie's preferred red for their octagon saleroom. Wilson enjoyed technology, and with the globalisation of the art market he introduced electronic conversion boards into the Sotheby's saleroom. A significant innovation in the conduct of sales was the installation of telephone bidding, which gave buyers complete anonymity and relieved them of the need to attend a sale in person. Despite its convenience, it had the effect of slowing down bidding, and some of the dramatic pace was lost.

Above all, Sotheby's supremacy was based on its success attracting American customers via its New York office. John Carter, its first representative in the city, used the argument that a Cézanne will bring the same $100,000 hammer price whether it is sold in New York, London or Paris. The important point was what the vendor put in his pocket: in New York that would be about $75,000, in Paris after commission and taxes $65,000, but in London the client would pocket $90,000. It was a convincing sales pitch and by the 1956–7 season, American consignments were responsible for more than 20 per cent of Sotheby's turnover and rising.

The most momentous change since Goldschmidt was Sotheby's takeover of the venerable New York auction house, Parke-Bernet.* The sale of the firm had generated great interest from both Sotheby's and Christie's, and a consortium of French auction houses entered the negotiations. The problem was Parke-Bernet's insistence on being on equal terms with any new owner, a requirement Peter Wilson described as creating a dragon with two heads. Sotheby's was already much larger and could not accept this condition. Parke-Bernet was housed in a custom-built auction house with a killer lease, an annual rental of £85,000 plus a further 2 per cent of *gross income* once turnover reached $6 million. Only a new owner could renegotiate the terms of such an onerous agreement.

* This was precipitated by the death of Parke-Bernet's English-born chief executive, Leslie Hyam, in 1963.

The 1964 talks with Sotheby's were complicated and protracted. A telling moment came when Wilson surprised his colleagues by asking to see the Parke-Bernet telephone bill. He had never shown any interest in company expenses before and in fact, he had the opposite purpose. 'They never make calls!', he said triumphantly, confirming his suspicion that the firm was not proactive and at once resolved to acquire it. The Sotheby's board was divided about the acquisition. Some, like Anthony Hobson, thought it was foolhardy and risky, and that New York would eventually turn round and eat them (he was right). But the younger people, headed by Wilson, were strongly in favour, as they saw the United States as a vast reservoir of business. At an impasse in a board meeting, Wilson turned to the most respected of the older mentors, Jim Kiddell, and said, 'What do you think Jim?' To which Kiddell replied, 'I believe the young people in the firm want it, and it's their future we should be thinking of ... I believe we should do it too.'[24]

With the Sotheby's purchase of Parke-Bernet, the American art world felt their heritage was threatened, and a hostile PR campaign ensued. One shareholder told the *New York Times* that the American flag was being sold down the river. The Parke-Bernet takeover was seen as a battle between the old Americans and the young British, and many of the old American staff left. Parke-Bernet was deficient in expertise, so Sotheby's London was trawled for young experts. Most of them were very young indeed, including the dashing English boss, Peregrine Pollen, already successfully running the existing Sotheby's New York office.

Behind the gleaming façade of Sotheby's success was a chaotic administration, with every department doing things its own way. Although departments remained independent fiefdoms until well into the 1980s, it was Wilson's rash decision-making that would cause the most problems. Providing impressionist buyers with credit advances on sale arrangements, he had concealed these payments from the partners. Such arrangements worked in a rising market, but when the market faltered in 1969 it emerged that an English art dealer in Paris, Stephen Higgons, was unable to service a debt of £400,000. This caused anger among Wilson's colleagues, and such rashness combined with over-rapid expansion left Sotheby's with a huge hole in their finances. With many payments outstanding from buyers, the firm went into financial crisis, its overdraft soaring to £1 million.

At this critical juncture, an offer appeared that seemed to Wilson like a godsend. In 1971, the tobacco company WD & HO Wills was looking for a new brand name for a cigarette and market research suggested either Sotheby's or Fortnum & Mason. Not only was £100,000 being offered, but also a great deal of free publicity and a commission on sales. The board was sharply divided, and some directors immediately resigned, but Wilson pushed it through. The cigarette affair became a vote of confidence in his management, and he demanded loyalty. It was only the commercial failure of the 'Sotheby's' cigarette – by all accounts disgusting – that averted more resignations. Wilson had made a serious misjudgement, which damaged his own reputation and that of the firm. According to Adrian Eeles, 'Peter never forgot who had opposed him', and 'This error of judgement completely changed the character of the firm.'[25]

Chapter 3

The Christie's Fightback

'The most important part of the recovery was Christie's ability, in the early 1960s, to hold onto its old clients, to keep them and their families faithful.'[1]

Lord Hindlip

If the 1960s were Sotheby's years of giddy expansion, by mid-decade Christie's was starting to fight back on the territory it knew best, old masters and fine British furniture. As the dealer Richard Green pointed out, despite Wilson's promotion of impressionist and modern painting, the day-to-day in London was still dominated by old masters and British paintings, which played to Christie's strengths.[2] The most important collection to be put up for auction since the war was the property of the late Captain George Spencer-Churchill from Northwick Park, sold in 1964. It was a fascinating mixture of Ancient Egyptian art, old master paintings and what Spencer-Churchill called 'the Northwick Rescues', paintings that he had restored back to life. He had always liked to ascertain from visitors: 'Are you a BC or AD person?'* The Christie's sale attracted great interest,

* Captain Spencer-Churchill had been wounded in the head, which left him a bit strained, according to Paul Levi: 'His habits were so odd that he had the dining room wired and if somebody upset him he would turn up the music, Wagner, very loud.' Supper might consist of nothing more than a boiled egg.

but the executors were so averse to publicity that John Herbert found it necessary to run round to Hatchards to buy twenty copies of *Great Private Collections* (1963), which contained the only description of the collection, to give to journalists. The best paintings and antiquities were sold in London for almost £1 million and the house auction totalled £68,597.

The Northwick Park sale gained notoriety owing to the *Sunday Times*'s exposure of a dealers' art ring. Colin Simpson, an investigative journalist, had posed as a member of the ring and managed to get into the post-auction 'knock-out' at the Swan Inn at Moreton-in-Marsh. He wore a concealed microphone which transmitted to a tape recorder in a van parked outside. Of the fifty dealers present, three were members of the council of BADA, including Major Michael Brett, who, along with twelve others, resigned.[3] The result was that BADA would make members sign a declaration that they would not take part in rings, although this did not prevent them from doing so, as we shall see. As far as Peter Chance was concerned, this was a country auction problem: 'There's nothing of the sort operating in London. Too many people chasing too few goods – you can only have a ring operating effectively if you've got everyone in it. It is a very different affair in the country because there you have the country dealers all knowing each other.'[4] The *Sunday Times* article about the Northwick Park ring even led to questions being asked in parliament.

If Christie's had the preponderance of grand British clients, the main battleground was over impressionist paintings in America. They did not achieve parity with Sotheby's until the 1980s, but the foundations were laid early. Christie's had opened a representative office in New York in 1959, but what changed their fortunes in America was the arrival of David Bathurst in the London office in 1963. The old master specialist David Carritt had suggested that Christie's poach Bathurst (Eton and Oxford) from the Marlborough Gallery, and he was the first person to give Sotheby's a run for their money in the modern field. Reckless, clever and engaging, he had more flair than his Sotheby's rival, Michel Strauss.

Making frequent trips to America, Bathurst scored the first major consignment with Monet's *Terrasse à Sainte-Adresse.* It was owned by the Reverend Theodore Pitcairn, who hung it in his pantry at Bryn Athyn, Pennsylvania. Pitcairn had bought the painting in New York in 1926. Wandering down East 57th Street, he had seen it in a dealer's window, and within ten minutes paid $11,000 for it. Pitcairn had founded his own church, the Lord's New Church, which was a splinter of the Swedenborgian Church. He was a talented art collector, buying paintings by Van Gogh in the 1920s. In 1967, he told Bathurst that he wanted to sell the picture and invited him to spend the night. Whilst there, Bathurst surreptitiously glanced at his host's diary open on a desk and saw the ominous initials 'DW' on one day and 'PW' for another; these obviously signified Daniel Wildenstein, the richest and most secretive dealer in the world, and Peter Wilson. Bathurst realised he had to make his proposal and win the sale before the two goliaths came onto the pitch. He succeeded, and *Terrasse* arrived in London for sale at Christie's in December. CBS sent a team to make a film about the city as the 'centre of the international art market'. The painting sold for 560,000 guineas, the highest price for an impressionist painting sold in Europe. It was bought by Agnew's on behalf of the Metropolitan Museum in New York.

For anyone entering Christie's headquarters in King Street, the first person they would probably encounter was the ebullient Ridley Cromwell Leadbeater, who managed the front counter. After thirty years, he knew all the customers and more about what was going on than anyone else in the saleroom. The Victorian Department director Christopher Wood described

Leadbeater as 'a mixture of club hall porter and family butler. Always immaculate, Mr Leadbeater sported a rose in his buttonhole ... it was his job to show us the ropes, as a sergeant-major might instruct a young subaltern or the gamekeeper teach the young squire to shoot.'[5] He was full of cockney stories and songs, mostly of the 'petticoats up and trousers down' variety. Often asked by clients to bid on their behalf, Leadbeater would be given a generous commission, with which he would treat the younger auctioneers to a slap-up lunch at Frank's, where he was well known and they made a fuss of him. The front-counter juniors were like boarding school fags 'sent on errands by the directors, parking their cars or buying smoked salmon from Fortnum & Mason'.[6] According to the trainee Dick Kingzett, who would later become an Agnew's director, the first task of juniors on Christie's front counter was to open the post each morning. They were advised that anything with a foreign stamp should be put in the bin, as 'it only means trouble'.

To break this old-fashioned way of operating required the determination of Guy Hannen, the single most important figure in Christie's modernisation. The staff were rather frightened of the all-seeing disciplinarian and details man, who, as one colleague put it, was a company mechanic. Hannen had won an MC in the war and had a military approach to staff management, telling a colleague: 'you have to be a cunt or a bastard to run people'.[7] He held the common misconception that someone from outside was always better than somebody already inside. This often led to disappointment, but he was absolutely right to appoint David Carritt, the most brilliant old master expert in the art trade. Hannen's managerial approach had the perfect foil in the new chairman, the avuncular Jo Floyd, appointed in 1972. Universally loved and respected, Floyd and Hannen were yin and yang.

In the 1960s and early 1970s, Christie's preeminent Old Masters Department held a series of spectacular sales only matched by the colourful characters it nurtured. The central figure in the department was the Hon. Patrick Lindsay, 'a dashing figure, with bushy eyebrows like permanent exclamation marks'. One of his colleagues recalled that, 'as a result of his passion for vintage cars and aeroplanes, he was often in plaster or a wheelchair and had to be pushed around by his invariably pretty assistants'.[8] There was nothing Lindsay enjoyed more than landing his Spitfire on the lawn to take a country house sale. He had the reputation of being very amusing, difficult to work with and strong-willed, but capable of an apology and even an invitation to lunch. In the opinion of Christopher Wood, 'he had no time for clients and could be astonishingly rude and off-hand to them, unless they were Dukes'.[9] Lindsay knew all the great families and their treasures and was the natural auction house figure to consult when a Titian needed to be sold to pay death duties or mend the roof. Owing to Lindsay's administrative deficiencies, Guy Hannen was made overlord of the picture department, which Lindsay resented, and a power-struggle ensued, making picture meetings rather tense.

To improve their expertise, in 1957 the firm hired a Courtauld graduate, Brian Sewell, on the recommendation of Anthony Blunt. Alec Martin's first words to Sewell were more accurate and wounding than he knew: 'Well, we know you've no social connections.'[10] Sewell was to write an amusing but tendentious memoir of his time at Christie's, dwelling on homophobic slights from Patrick Lindsay, of whom he painted an absurdist portrait as somebody who believed aristocratic ownership to be a proof of authenticity.* In his eccentric way, Sewell was an effective expert and identified an El Greco which, to his fury, his colleagues refused to accept. Spotted by Agnew's, it was sold to Charles and Jayne Wrightsman.

* Sewell, who liked cars, noted enviously that 'Patrick drove an Aston Martin DB2, Jo Floyd a Jaguar 2.4 and Peter Chance an R-type Bentley.'

Christopher Wood described Brian Sewell as small, curly-haired and with an extraordinarily plummy upper-class accent. He wrote that working with Brian was difficult: 'not only was there his voice but he was also prone to tantrums when the voice would rise to a suppressed scream. One day Brian became so impossible that Noël Annesley and Charlie Allsopp dragged him from his chair, carried him down the stairs, across the hall, and deposited him on the pavement outside. "Naughty children" said Brian, who clearly enjoyed it.'[11] Brian, with his Courtauld background, was appalled to find himself cataloguing so much low-value nineteenth-century trade fodder. Sewell could equal Lindsay in his rudeness to clients, telling one old lady what to do with her horrid painting: 'If I were you, Madam, I should take it out into King Street and stamp on it.'[12] The correct procedure for items that were beneath Christie's threshold of value was to send them to Bonhams: '*Nil nisi Bonham*,' as one director quipped.*

In-house expertise was bolstered by a regular traffic of museum experts, who would do the rounds of Sotheby's and Christie's: Teddy Archibald of the Maritime Museum in Greenwich, David Piper of the National Portrait Gallery, and old masters were sometimes submitted to Ellis Waterhouse or Michael Jaffé. One oft-repeated old chestnut involved several Christie's experts looking at a nineteenth-century English landscape painting and having no idea who it was by. 'It's a real bastard', said one, and thus the artist became 'Bastard'. Somebody shouted, 'Give me a Christian name!' 'Lawrence', said another, and thus it went into the catalogue. There was a darker side to the job, according to Sewell, who recalled that dealers would offer him a fiver to ensure that pictures were included in certain sales, change the placing in the catalogue or tweak the attribution. As Sewell

* The Latin phrase *De mortuis nil nisi bonum dicendum est*, often abbreviated to *Nil nisi bonum*, meaning 'Of the dead nothing but good is to be said'.

discovered, 'behind Christie's grand façade ran two economies – the upstairs theatre of the sale room with a suave director playing auctioneer, and downstairs the black economy of staff so poorly paid that they depended on the tips and bribes that came their way'.[13] There is no doubt that Sotheby's and other auction houses were similar in this respect.

Sewell's nose would be put out of joint by the arrival in the Old Master Department of David Carritt, the art world *wunderkind*, in 1963. Carritt's discoveries were legendary: a set of Francesco Guardis in Ireland, a Caravaggio in Cumberland (*The Concert*, now in the Met in New York) and a Dürer in a Norfolk country house. Carritt brought energy and brilliance to Christie's. As Christopher Wood noted, 'We young ones at Christie's much admired Peter Wilson and wished we could have him for our chairman. We thought our own directors too fuddy-duddy and old fashioned. That is why David Carritt was so vital to Christie's, as he literally turned things around, by his reputation and his uncanny knack of finding things.'[14] When Wilson was informed about Carritt's appointment at Christie's he commented drily: 'It is the first example I've heard of a rat *joining* a sinking ship.' Carritt was himself devastatingly witty. Wood described him as looking 'florid, pink and a little bland until animated by talk and laughter. Always neatly dressed and brushed, with a coif of hair pushed back; he walked mincingly and was extremely camp. He lived in Mount Street in a flat below the theatre critic and flagellant, Kenneth Tynan, where he claimed to hear howls and sounds of whipping coming from above, "just a thong at twilight my dears".'[15] According to Wood, 'David and Brian would often engage in vicious cat fights'.*

* Sewell was overlooked for promotion and left, describing Lindsay as 'an incubus'. In 1970, Carritt announced he was leaving Christie's, lured away by Baron Lambert to set up Artemis, discussed in the next chapter.

The years 1964 and 1965 turned out to be bumper years at Christie's, their turnover doubling thanks to a succession of great sales. In addition to Northwick Park, there were the treasures from Harewood House and the celebrated Rembrandt portrait of his son Titus, belonging to the Cook family. The story of the sale of *Titus* was troubled from the start. Sir Francis Cook insisted on the picture being flown in a Dakota from Jersey to Gatwick. Christie's had to fight off the offer of a guarantee of £500,000 from Peter Wilson. Sir Francis challenged Peter Chance with 'You can do better than that, can't you?'[16] Chance accepted the challenge with considerable anxiety. It was believed that Sir Francis had already offered the picture to Norton Simon via Agnew's for £650,000. It was recognised that Simon was likely to be a bidder and had already ordered a seat in the middle of the main saleroom. The farce of the sale began with Charles Allsopp, the firm's young rising star, being told to hire a Rolls and meet Simon at Heathrow with a copy of the *Financial Times*, a white carnation and a pair of dark glasses. When Simon outlined his plan for bidding, Peter Chance told him it was too complicated and he should get Dudley Tooth (his dealer) to bid for him. Simon would have none of it and his bidding plan revealed itself in horrifying public pandemonium on the day.

When the bidding reached 740,000 guineas (Christie's had not yet converted to pounds), Chance turned to Simon and repeated several times 'against you'. He then brought the gavel down, announcing 'Marlborough Fine Art'. This was greeted by applause until Simon stood up shouting 'Hold it, hold it', to which Chance answered, 'No I said quite clearly against you for the last time.' This caused a furore as the room erupted at the astonishing impasse. Nobody knew what was happening, but the television cameras were rolling. Chance had to rap his gavel to restore order. Norton Simon then asked Dudley Tooth to publicly reveal his instructions: 'Alright I will read what's in my pocket … When Mr. Simon is sitting down he is bidding. When he stands up, he has stopped bidding. If he then sits down again, he is not bidding until he raises a finger. Having raised his finger, he is continuing to bid until he stands up again.'[17] In light of the instructions, Chance, although furious, had no option but to restart the bidding. This was disputed by David Somerset of Marlborough, but Chance responded: 'I am afraid my decision must be final. It's laid down clearly in the Conditions of Sale, if there is a dispute about a bid there is no alternative but to put the lot back into sale and continue with the bidding.'[18] This time, the bidding went to 760,000 guineas, which was the highest price for any painting sold in Britain. Afterwards, Peter Chance left the room 'As grim as a bulldog about to attack and very red in the face'.[19]

There was a furore in the press and Geoffrey Agnew wrote to *The Times* complaining about the behaviour of accepting such impossible instructions. The sale held one further punishment for Christie's. Sir Francis demanded to be paid while the picture was delayed for an export licence procedure and the firm lost the considerable bank interest charges. This was all very much against Peter Chance's favourite dictum, 'Get it right.' As the young art dealer Giuseppe Eskenazi commented on the *Titus* affair: 'The squabbling millionaires had injected a moment of raw drama into a world always keen to project itself as a bastion of calm, effortless, double-breasted gentlemanly assurance.'[20]

David Carritt proved his worth to Christie's when he discovered a Tiepolo in the ceiling of the Egyptian embassy at 75 South Audley Street, where the *Allegory of Venus Entrusting Eros to Chronos* decorated the drawing room. Carritt was browsing through the Morassi catalogue of Tiepolo's paintings and found some tantalising clues. He was no doubt stimulated by reading the present 'whereabouts unknown'. Making the connection with the previous owners of the Egyptian embassy, he sought permission to look for it and secured the sale. The Tiepolo made £409,500 in 1969 and was acquired by the London National Gallery. That season, Christie's had a worldwide total of £11.1 million, which was still far behind Sotheby's £40.3 million.[21]

Although he was still two years off retirement, in 1970 Chance relinquished the rostrum in favour of Patrick Lindsay for the sale of the Earl of Radnor's Velázquez *Portrait of Juan de Pareja* from Longford Castle. The work was expected to be the first picture to break the million-pound barrier. On 27 November 1970, the bidding started at 250,000 guineas. Christie's director, William Mostyn-Owen, was bidding on behalf of a private collector on the telephone but dropped out at the million mark. The dealer Hugh Leggatt came in briefly, but the auction developed into a duel between Geoffrey Agnew and a man in the second row, who took the bidding to 2.2 million guineas. Looking at Geoffrey Agnew, Lindsay offered 'One more?' but it was not to be. Agnew's had been so confident of acquiring the painting that their van was parked outside. Nobody knew who the winning bidder was, and Lindsay was not saying. It was in fact Alec, the son of Daniel Wildenstein. Alec had bid on behalf of the Metropolitan Museum in New York and had been instructed to buy it whatever the cost. The price was far in advance of anything previously achieved, and it set a new benchmark for works of art at auction. The Velázquez record would stand for ten years.

For Colin Simpson writing in the *Sunday Times*, it marked the end of an era in the salerooms and of the Christie's employee who simply had a good address book and a few years' service in the Guards: 'Friday's sale sounded the death-knell of this concept, because it was a triumph for the intellectual midriff of the corporate animal Christie. The painstaking four-page scholarly catalogue entry was the work of David Carritt and William

Mostyn-Owen. The bibliography was one of the most detailed ever produced by the firm.'*[22]

Sold in 1971, Titian's *Death of Actaeon* was the third in a trinity of masterpieces handled by Christie's. It had been on loan to the National Gallery from its owner, the Earl of Harewood. The first public inkling of a sale came when a lady went up for the day from Budleigh Salterton to see the painting and wrote to *The Times* complaining that it was not on view. When Christie's realised that the story was out, Johnny van Haeften, the Press Office junior, was instructed to drive round Fleet Street handing out a press release. Perhaps rashly, he started with *The Times*, who, feeling proprietorial over the story, kidnapped van Haeften to prevent him giving it to the other newspapers. The painting was bought by the maverick dealer Julius Weitzner for 1.6 million guineas and is now in the London National Gallery.

When Christie's chairman, Lord Hindlip, later wrote about the firm's recovery in the 1960s, he was right in maintaining that the loyalty of old clients was crucial, but he added that this was 'not as some of our colleagues believed, because Christie's had a God-given right to sell certain families' works of art, but to ensure that Christie's would still be consulted'.[23] One of the advantages of loyal clients is that they are less likely to negotiate on terms, and this greatly helped Christie's balance sheet. Their profitability enabled them to go public as a company before Sotheby's. Peter Wilson's success was still dazzling on the surface. During the 1972–3 season, Sotheby's sold £72 million against Christie's £34 million. Christie's profits, however, were just over £1.1 million, and Sotheby's, despite the much higher turnover, were fairly similar at £1.4 million.[24] Christie's went

* Christie's did not have it all their own way – occasionally, Sotheby's had sensational old masters like the Rubens *Adoration of the Magi*, now at King's College Cambridge, which they sold in June 1959 for £275,000, a record for a picture at auction at the time. In 1961, Sotheby's also sold the notorious Goya of the Duke of Wellington, now in the National Gallery, for £140,000.

public in 1973, and the stock was eleven times oversubscribed. Their timing was fortunate, as the economy was hit later that year by the oil crisis.

The success of Christie's public offering and the concomitant windfall to their partners was not lost on the directors of Sotheby's, but it was clear that the firm needed more financial discipline before this was possible. Wilson's answer was to bring in Peter Spira from Warburg's to be the finance director in 1974. It was to be painful on both sides. When somebody asked Spira why he had left the City for such a 'hot potato' as Sotheby's, he admitted that there might have been an element of naivety in the decision, 'because I didn't realise Sotheby's *was* such a hot potato'.[25] Spira immediately went to war on the directors' expenses. He put a considerable dampener on risk-taking, and the gambling instincts that tried to win sales at any cost. One of his most controversial edicts was to demand that the art trade must pay their bills within a month (when Sotheby's paid the vendors) as opposed to the three or four months' leeway they had become used to. Geoffrey Agnew protested to Wilson, who stormed into Spira's office: 'I've been in this company for 40 years. You've been here four weeks and you're ruining the business.'[26] As one director recalled, these were the years 'when the meetings started'.[27]

For the first time, profit rather than growth became the focus of attention at Sotheby's. The main problem was that the commission rates, now averaging 14 per cent, could hardly be increased if they were to maintain their competitive edge. By 1975, inflation was running at 20 per cent, and both auction houses saw their profits plunge. As a result, they decided to introduce the hugely controversial buyer's premium. For the dealers, the buyer's premium was the Big Bang of the art world, giving the auctioneers a decisive competitive advantage. The question is who first thought this up and was there collusion between the two houses? Anthony du Boulay, Christie's porcelain expert, remarked that he had lost several ceramic collections to Continental auctioneers (particularly Swiss) because they were charging a buyer's premium and could therefore offer a lower seller's rate: 'I thought we needed to follow their example.'[28] Christie's decided to adopt this practice. Christopher Wood had no doubt that the buyer's premium was one of the results of the 1970s recession, noting in his diary on 31 May 1975: 'There had been frequent conspiratorial meetings recently between Jo Floyd and David Westmorland [at Sotheby's]. We, of course, pretend to know nothing of this. "Collusion? God bless my soul, no!"'[29] However, 'unlike the later collusion scandal of the 1990s, it was all done very discreetly, and with no paperwork of any kind'.[30]

On Friday 30 May 1975, Jo Floyd formally telephoned Sotheby's to inform that his firm would be announcing the introduction of a buyer's premium the following day.* To many, this was the end of art world fraternity, and especially the end of cordial relations between dealers and the salerooms. From now on, it was 'them and us'. Relationships with the dealers were patched up but never fully repaired. Phillips saw their chance and announced they would reduce their commission rate to 10 per cent and charge no buyer's premium. This gave them short-term goodwill in the trade, but three years later they followed suit as they found it was no longer possible to survive economically without a buyer's premium. The trade was inevitably incensed by the whole idea, not just because of the extra cost to themselves, but also because of the competitive advantage it gave the auctioneers, along with a general sense that it was not playing fair. In the opinion of most dealers, it had two damaging results: auctioneers usurped their role – one for which

* They did not apply the premium to Christie's South Kensington, where the seller's commission remained at 15 per cent. Jo Floyd had written to the Dealer's Associations and tried to sweeten the pill by doubling the normal introductory commission to 4 per cent.

they were quite unsuited – of nurturing buyers, and they put themselves into an unfairly competitive position from which they could reduce, and sometimes waive, commissions from the seller. Julian Agnew, like many other dealers, felt a keen resentment towards Sotheby's and Christie's, and Wilson in particular.

The immediate effect was a boycott of the first Sotheby's sale in the autumn, a silver sale. Dealers, however, were not as united or determined as they may have seemed and were quietly finding alternative arrangements for bidding. In New York, Christie's imposed the buyer's premium immediately, which Sotheby's New York did not. They believed, like Phillips, that this would buy goodwill, but it was dramatically offset by Christie's ability to reduce terms and win business.

When the premium was announced, Hugh Leggatt, who had a reputation for pomposity where the auction houses were concerned, exclaimed, 'I was first of all amazed, then I was absolutely *furious.*' He went round to see Jo Floyd, with whom he had an extremely disagreeable meeting, and then with Peter Wilson, which he described as *vicious.* 'I called him every name under the sun. I wasn't as rude to Jo Floyd, but it was quite useless to try and budge him. They both said they needed the money, I said I thought it was absolutely appalling.'[31] There was a big debate about what to do at SLAD. The chairman, John Baskett, went to see Lord Borrie, then director-general of the Office of Fair Trading, whom he describes as 'of a left-wing persuasion. When I pointed out to him that, at auction, the buyers' interests were quite the opposite from those of the seller and that it was illegal on this account for Estate Agents to charge both parties, he replied, "Ah, but yours is a luxury trade, whereas everyone needs a roof over their heads."'[32]

Martin Summers of Lefevre Gallery commented:

> The buyer's premium came as a great shock. The dealers had no doubt that collusion had taken place. They believed that

> probably at a mid-level the information was passed from one auction house to another, then debated at boardroom level and announced by both houses almost simultaneously. The dealers took legal advice but realised they had no case, so instead they went for maximum publicity and embarrassment which brought the auction house to the table.[33]

With arguments pinned on the charge of collusion, the battle between the dealers and the auction houses raged for the next six years. It was eventually settled in 1981 in a deal sealed at Claridge's hotel, where each side had taken a suite. Floyd reported to his board in September that the previous evening there had been a meeting in Claridge's, finishing at 3 a.m., between Christie's and Sotheby's and the two main dealers' associations. The dealers announced that 'As a gesture of goodwill and in the interests of preserving London's position in the international art market', the plaintiffs had decided to withdraw their action and the auction houses agreed to pay half of the dealers' costs, which amounted to £75,000. It was also agreed that, in future, prices would be quoted including the premium and differentiated from the hammer price without it. Relations between the two sides would take a generation to heal, although good personal relationships remained with specialist departments.

The introduction of the premium improved the balance sheet of Sotheby's to such an extent that by 1977, Peter Spira could recommend to the board that Sotheby's should go public. He forecast a profit that year of £4.6 million, and the following year the pre-tax profit was over £7 million. The Sotheby's IPO was oversubscribed twenty-six times. The three largest shareholders were Peter Wilson, Peregrine Pollen (then seen as Wilson's heir) and Jacob Rothschild (the Rothschild Investment Trust had bought Anthony Hobson's partnership when he resigned).

The introduction of the buyer's premium was a watershed

moment for the art trade. Going public made the partners rich men but also meant that the two firms had lost control over their own long-term future and would be vulnerable to take-over. In the short term, this hardly mattered.

Chapter 4

The Grandees

'Well-established firms, especially old family businesses like Agnew's, Colnaghi and Leggatt Brothers, give the appearance that they will go on forever.'[1]

Barrie Stuart-Penrose

The horrified opposition to the expansionist auction houses included the old family Bond Street dealers who had traditionally controlled the international market. The grandest of them was Agnew's, which had a proprietorial attitude to the British market. No dealership in London apart from Leggatt's and Colnaghi was more dependent on a 'gentleman' culture, but Agnew's, founded in 1817, was by far the most formidable operation. The chairman, Sir Geoffrey Agnew (1908–1986), sat in the front row of every major old master sale, flanked by his partners, as if by hereditary right. There was nothing nonchalant about them as they surveyed the room with a beady eye to see who was bidding. Before the war, Agnew's expected to be offered major paintings for sale ahead of the auction houses. In the new post-war reality, the firm repositioned itself as the natural alternative to Sotheby's and Christie's for a second opinion and as the preferred bidding agent at auction for the world's greatest museums.

Agnew's was in a different league from most London dealers

because it had the most impressive stock and the freehold of its Bond Street building with the grandest showrooms. An astute, commanding, bluff figure, and a bit of a monster to boot, Geoffrey Agnew effortlessly led the trade, ruling by force of personality. Admiring a picture in front of Sir Geoffrey was taken as a sale. He was a knowing man for whom early intelligence of a potential sale was vital. Part of the plan of staying in touch with owners was to stage exhibitions of their treasures, and these came from the Spencer, Devonshire, Lansdowne and Cholmondeley collections, amongst others. Agnew's put on impressive shows, but the gallery was an intimidating place to walk into.

To Lord Hindlip, chairman of Christie's, 'The trouble with Geoffrey and Agnew's in general is that they preferred knighthoods, prestige and publicity – they never made a profit and were not interested in business.'[2] Geoffrey Agnew, an Old Etonian who had once taught history there to his future clients, was shrewder than that analysis suggests. The firm had an impressive array of knowledgeable directors (most of them family members and their children), particularly Evelyn Joll, the acknowledged Turner expert, and the popular Dick Kingzett, who had a fine track record of selling British art to German museums. Its expertise allowed the firm to have departments for paintings, drawings, prints and sculpture.

As impressive as their expertise was their premises, with its yards of red damask and dark oak, characterised by the art critic John Russell thus: 'At Agnew's the visitor is soon aware of that perennial element in English life: the country house connection. The rooms and staircase could be part of some vast remodelled mansion, with a gun-room somewhere off the hall and huge dogs, still warm from the chase, momentarily absent from the scene. Frayed red damask sets the tone of a family too grand to bother with appearances.'[3] If you got past the hall and were deemed rich enough (or serious enough), you might be lucky to make the inner sanctum, which was fairly similar to other grand dealerships. Dick Kingzett's son Christopher, who also worked at the firm, described the Agnew's sanctum as resembling something out of *The Wizard of Oz*: 'at the touch of a button, the heavy velvet curtains that at first glance seemed to be a wall-covering drew aside to reveal a painting. Two more buttons only needed to be touched for the curtains to slide back on the side walls of the room and two more paintings were revealed in all their glory.' Kingzett noted, 'At any point in time, the three most important paintings at Agnew's were behind these curtains … It was not at all unusual for our regular clients to come in and say, "Let's see what's on behind the curtains in the private viewing room."'[4]

Agnew's held museum-quality stock and was the place to go to buy a Titian, a Rembrandt or a Turner. The business model had served the company well: purchasing from clients, holding stock, always thinking of the long term and doing 'the right thing' as they saw it. One of the directors reflected that, 'My family has lived for 150 years largely by selling pictures to one generation and buying them back from the next.'[5] Agnew's had capital and didn't have to take things on consignment, which gave them an advantage. The directors made a special point of drawing in smaller collectors, with their long-established exhibitions of watercolours at Christmas including items as cheap as £15. In this way, by the 1960s, they had expanded their client base to three times what it was in 1939.[6] Agnew's created a client loyalty. 'Isn't it strange,' a rival dealer noted for his malice once remarked, 'how some people remain faithful to Agnew's all their collecting lives. They don't even seem to enjoy buying a picture from anyone else.'[7]

Geoffrey Agnew pinpointed the difference between himself and contemporary art dealers: 'Unlike the experience of many modern dealers, our major difficulty is not selling works of art but finding them.'[8] Agnew's conspicuously did not sell impressionists 'when they were cheap', as Christopher Kingzett lamented, 'we thought they were bad and then they were out of our range'.[9] Agnew's tried selling Modern British art but never with much conviction. Julian Agnew once said to a colleague: 'The only decent artist is a dead artist', and although this may have been said in jest it represented a truth that they were more comfortable with the past. For Kingzett: 'You can't sell what you don't like yourself' – an idea Frank Lloyd at the Marlborough Gallery would have scoffed at. A director of Agnew's recalls: 'At one point we had a show of terrific work, a man who painted all kinds of sex with his girlfriend, a sort of sexual marathon; but it just offended too many people.'[10] It was not for their client base.

Agnew's focused on American buyers and Sir Geoffrey crossed the Atlantic fifty-six times between 1947 and 1967.[11] The biggest of them was Paul Mellon, to whom he sold – or 'placed', as he would have put it – seven oils and twenty-three watercolours by Turner, along with major paintings by Reynolds, Constable, Zoffany and Stubbs. In 1960 alone, Mellon bought forty-four paintings from the firm. Mellon once complained to Agnew that he had not received any photographs lately, because 'I do like to have a packet to open at breakfast, you know'.[12] The three illustrated volumes of the firm's history covering the period of this book attest to the extraordinary number of paintings sold to American museums, including such haunting masterpieces as Salvator Rosa's *Lucrezia as Poetry*, which went to the Wadsworth Atheneum in 1955.

The first shockwave to disturb Agnew's monopoly over selling masterpieces for grand families and acting on behalf of museums was the sale of the Earl of Radnor's Velázquez *Portrait of Juan de Pareja* at Christie's in 1970. Julian Agnew saw it as a pivotal moment. The firm was not consulted over the sale and to add to the humiliation they were outbid at the auction by Wildenstein's, operating on behalf of the Metropolitan Museum in New York.* Until then the most important old masters were generally sold privately through dealers. Some of the younger members of the trade, like Richard Green, regarded the auction houses as colleagues to negotiate credit terms with, but to Agnew's they were deadly competitors. They would, however, occasionally cooperate with other dealers such as Marlborough, whose Japanese clients wanted paintings by Turner. According to one director, Agnew's worked with Richard Green until they found that he was making more money out of the deals than they were. The great battleship that was Agnew's would sail on to have its notionally best year as late as 1989, but time was running out and the business model was to be fatally undermined.

'Colnaghi, Agnew's, the Heim Gallery and Hazlitt's, between them dominated the old master field,' wrote Jeremy Howard, the historian of Colnaghi, 'but where Hazlitt would have the sketch, Colnaghi often had the altarpiece and while the ethos of Agnew's was redolent of the English country house, the atmosphere of Colnaghi's was more that of an Italian palazzo.'[13] In Christopher Wood's view, Colnaghi's had a greater reputation than Agnew's during the 1960s. Even the acerbic Brian Sewell described Colnaghi as 'the only art dealers in London of never questioned respectability'.[14] Its reputation rested on its illustrious history and its scholarly partners, including two of the most distinguished tastemakers in the art trade: James Byam Shaw and Roddy Thesiger.

In post-war London, Colnaghi was always more modest than Agnew's despite an unrivalled reputation among curators, art lovers and collectors for discrimination and knowledge. It was the trade equivalent of the Ashmolean or Fitzwilliam Museum. Founded in 1760, Colnaghi's had its golden age between 1894 and the 1930s.[15] The firm had survived the war, uncertain each

* One story has it that Wildenstein had been trying to buy the Longford Castle Velázquez for some time.

morning whether the building at 14 Old Bond Street, where it occupied five stories, would still be standing. The first floor contained the main picture gallery, and the second the 'killing room', elegantly decorated with chairs arranged in front of a curtain which would be drawn back to reveal a picture on an easel. Prints and drawings formed a major component of the business and these were kept in solander boxes on the fourth floor. On the top floor was a framing, mounting and cleaning department. The third floor was administration.

Specialising in drawings, James Byam Shaw, always known as Jim, was the ideal of the scholar-dealer. With a family connection to the firm, it was assumed that, after Oxford, he would join Colnaghi. Before doing so he was given an astonishing seven years to educate himself in the print rooms of Europe, recording everything he saw in notebooks now preserved at the Ashmolean. To serious drawings collectors, Byam Shaw was the most highly regarded figure in the trade. It was said that nobody could disagree so gracefully over an attribution. You left his company believing that the art world was a good place. Such was his reputation for finding 'sleepers' and buying quality that he found it necessary to stoop to subterfuges in the saleroom. When he spotted an unrecognised work by Pieter Bruegel the elder at Sotheby's, the sharper members of the trade were watching him to see what he would do. He remained motionless while an unknown bidder at the back of the room – a porter from Colnaghi – secured the lot for the British Museum.

Possibly the most painful moment of his career was when he was brought in to value the entire collection of prints of the Albertina in Vienna, for sale to the Museum of Fine Arts in Boston. Byam Shaw found the situation 'rather awkward' but in the event it was narrowly averted. He always tended to put the academic and long term ahead of short-term commercial interests, happy to sell to a young collector for minimal profit. His greatest client was Count Antoine Seilern, and indeed he was almost the only art dealer admitted to the collector's house in Prince's Gate.

Roderic Thesiger, the firm's old master paintings partner, joined Colnaghi in the 1950s after war in the Welsh Guards and a spell at both the Tate Gallery and Sotheby's. Both Jim and Roddy had been to Christ Church, Oxford, and were almost alone at that time in the trade in having a university degree. According to John Baskett, a young staffer, there was tension between Eton-educated Roddy and Westminster-educated Jim. Thesiger's great achievement was his pioneering advocacy of Italian baroque or *seicento* paintings, for which the firm became celebrated. He was not alone in this, as we shall see, but he put Colnaghi at the centre of this developing taste. It is difficult today to understand the depth to which *seicento* and Bolognese painting had fallen in critical acclaim. A group of brilliant post-war art historians, notably Ellis Waterhouse, Nikolaus Pevsner and, above all, the scholar-collector Sir Denis Mahon, had started seriously to study the *seicento*.

As Julian Agnew put it, 'My generation was educated in *seicento* paintings.' Firms like Agnew's, Heim, Hazlitt, Julius Weitzner, David Koetser and Trafalgar Gallery would stock them, but to Colnaghi's, *seicento* was central.* Most of the paintings came from country houses, ending up at auction – notably items from the Ellesmere collection at Christie's in 1946 – but Roddy Thesiger had a coup in acquiring a group of pictures from the Barberini collection in Rome that had been given export licences. This taste would be transmitted to the next generation of the trade through others, including the dealers Michael Simpson and Patrick Matthiesen. When one looks at American museums' riches in this field, it reflects the

* During the 1970s, Agnew's, for instance, sold six major paintings by Guido Reni; the majority went to the USA.

fascination and scholarship of the London art dealers who supplied so many of them.

The specialist English school partner at Colnaghi's was Tom Baskett, on whose son, John, fortune smiled. When he took charge of British paintings and watercolours, John caught the eye of Paul Mellon, who had begun to specialise in British paintings. The art establishment tended to look down on British art, with the exception of Turner and Blake, and Mellon found that he had the field pretty much to himself. He wrote: 'I came over to London at various times of year, usually when my horses were running at Newmarket, Epsom or Ascot. As the knowledge of my collecting interests spread, I am told that on my arrival in the capital, the London dealers' windows filled overnight with English pictures.'[16]

John Baskett joined Colnaghi in 1957, as an assistant to his father, having spent five years studying at the British Museum and elsewhere. He appealed to Paul Mellon, who was shy himself. It soon emerged that Mellon only wanted to deal with Baskett, which made life slightly awkward. Thesiger ordered John to consult him about purchases for Mellon, and so he spent three years pretending not to be the great collector's agent. Finally, with a loan from Mellon, John Baskett left the firm to become independent and effectively his representative in London. Baskett was exactly what Mellon was looking for, an art dealer as understated as himself, who was knowledgeable, trustworthy and had good relations with all his colleagues in the trade. He was to be his patron's benign eyes and ears in London, playing a major role in the creation of Mellon's brainchild, the Yale Center for British Art. Mellon's way of operating was simple: he either accepted the price or rejected the picture and never bargained.

With the retirement of Jim Byam Shaw in 1968 and Roddy Thesiger in 1971, Colnaghi lost its footing, and although it continued to employ clever operatives it went through a succession of rich owners who never managed to recover its glory days. Jacob (later Lord) Rothschild was the first to acquire the firm. By his own account, he found Colnaghi 'rather troublesome: we made rather a mistake in recruiting the two brothers, Adrian and Nicholas Ward-Jackson. Adrian was like something out of a Raymond Chandler novel – he looked innocent but was actually lethal in all sorts of ways.'[17] Both brothers had hectic social and love lives and a destructive element which made it difficult to run the gallery. One day, Jacob Rothschild bought a painting from a young dealer, Richard Herner, who offered to pull the gallery round. 'Herner had a brilliant eye but was erratic and they had terrific rows,' as one colleague remembered. 'Rothschild always forgave Herner because he was so gifted, but they eventually fell out.'[18] The print specialist Adrian Eeles described his time at Colnaghi as 'being on a very rough sea'. Rothschild tried to expand the firm by inviting Michael Goedhuis to open a Persian and Indian miniatures department, but the firm didn't settle and there was a turnover of staff. As one former employee said, 'Jacob was always changing his mind and imposing his latest whim.'

When Rothschild subsequently joined the National Gallery Board of Trustees, he had to give up commercial activity in the art market. In 1980–1, he sold Colnaghi's to the German businessman Rudolf Oetker, who had been investing in their paintings, and his role gradually morphed into ownership. According to Richard Knight, who ran the firm after Herner, Oetker didn't have much interest in the art but was a relatively benign owner. Knight unashamedly wanted to put the clock back to the Thesiger years and specialised in filling gaps for American museums. The main problem he faced was that Oetker was against buying paintings with Christian subject matter, telling Richard to 'do it with your money, not with mine'.

Leggatt's were the third in the triumvirate of grand

old-fashioned dealers, albeit by far the smallest of the three, operating from a modest premises in Duke Street. The firm was a bit of a mystery. One dealer recalled a painting by Reinagle that seemed to be anchored in their window forever until swapped for a Bogdani. The patriarch, Hugh Leggatt, had a series of rich, distinguished, low-key clients like the collector Bobby Wills. Hugh's great love was for portraits: he was close to the National Portrait Gallery, to whom he presented his discovery of an unfinished head of Nelson by Beechey. 'His greatest role,' as Christopher Wood observed, 'was as writer of letters to *The Times*. He was literally tireless in writing about any art world matter, so much so that he virtually became the voice of the art world. For this, like Geoffrey Agnew, he received a knighthood, also for his services to the museum and art world.'[19] Leggatt was a great puller of strings behind the scenes and stood on his dignity with the auction houses, which he thought had got above themselves. He was understated and courteous; some thought he behaved as if he were 'head monitor' of the art world.[20] He never did fairs or exhibitions and depended on supplying the wants of a very small number of very rich customers, most notably the Thomson family of Canada.

The most mysterious of the great London family dealerships was Wildenstein, located in Nelson's old house at 147 Old Bond Street. Founded in 1875 in Paris, the firm has remained in the family ever since. Daniel took over running the business when his father died in 1963. His main interest in England was horses and racing. His London gallery manager, Max Harari, would be made to come in on Saturday and place the telephone next to the television set so that, wherever he was in the world, Daniel could listen to the racing. Wildenstein's main operations were the research foundation in Paris and the New York gallery. In the words of Jane Abdy, 'the London Gallery was the sleeping princess of the gallery world'.[21] When Julian Barran was Sotheby's impressionist auctioneer, he recalls, 'They were the only important dealer who never bothered to attend London sales.'[22] Adopting the motto 'keep and hold', the family held the greatest stock of art, had no need to buy at auction and showed little enthusiasm for selling anything. It was estimated that at one point they had 250 Picassos in store. Apart from Wildenstein's fabulous stock, it was the library and records which set them apart from all other dealers, and gave them an advantage in knowing where all the major paintings in the world were to be found.*

Nobody could recall anything Wildenstein sold through London – whether this amnesia reflects the actuality or the family's extreme secrecy is hard to tell.† Boucher tapestries hung in the entrance foyer and there was a reputed stock of Fragonards. Burton Fredericksen of the Getty Museum began to realise on trips to London that almost everything recommended by the art advisor and historian Frederico Zeri came from Wildenstein – although possibly not from the London stock – but it didn't seem to bother Paul Getty. Norton Simon, the trickiest customer of all, once said: 'When I buy at Wildenstein, I pay top price: but I pay for a top picture, and when I am ready to sell it on, I receive an even higher price back.'[23] Giuseppe Eskenazi once showed Daniel's son Guy around one of his Chinese art exhibitions, sighing that he had only sold half of the items, to which Wildenstein replied, 'That's brilliant! We only make money on the unsold paintings and not on sold ones!'[24]

* When the Royal Academy staged an exhibition of eighteenth-century French art in 1967–8, it could only be done with the willingness of the Wildenstein family, who loaned works from their stores, and put the organisers in contact with the collectors who owned them via Denys Sutton.

† John Lumley worked for Wildenstein briefly in the 1970s and confirms that they were doing little in London apart from running the family's racing affairs. In 1984, Earl Spencer sold Salvator Rosa's *Incantation of Witches* to Wildenstein for £200,000. The painting was stopped at export and bought by the National Gallery.

The man who inherited the mantle of Geoffrey Agnew and Hugh Leggatt and became the art trade knight was Jack Baer at Hazlitt Gallery. Baer projected a patrician image: elegant, amusing and clubbable. He had a considerable following in St James's, gaining the soubriquet 'gentleman Jack', although within the art world this tended to be repeated with a bemused irony. The great feather in Baer's cap was to become the chairman of the Acceptance-in-Lieu panel, a position of considerable power where the art trade meets museums and government in a benevolent embrace. Despite Baer's personal eminence, Hazlitt remained in the second league of dealers; it did, however, have one enviable client, the property tycoon Harry Hyams. As a dealership under Baer, Hazlitt sold baroque paintings to the Barber Institute in Birmingham, but where the gallery was ahead of taste was in promoting French nineteenth-century oil sketches.

The façade of these grand dealerships looked splendid, but the reality was that time was slowly running out for galleries holding a stock of important old masters. Increasingly, as the auction houses took over the market, the big deals were less about the stock held than what they purchased at auction on behalf of American museums.

Chapter 5

Bond Street and Beyond

'Very often the world of art fails to appreciate the contribution of influential art dealers in the formation of taste and collections, both private and public.'[1]

Tim Knox

Beneath the famous old names of Bond Street thrived scores of dealers trading in old master and British paintings. Many of them were clever émigrés. Typical was Lazar Herner, who arrived in London from Prague in 1938. On the boat, he met another penniless hopeful, Frank Schrecker. Both were mad about football and only knew one word in English: 'Arsenal'. After the war, they set up the Hallsborough Gallery and became partners for life. Frank had the ingenious idea that, while England was poor, recently neutral Sweden was prosperous by comparison, and he took two suitcases full of paintings over to Stockholm, sold them and returned time after time. Herner used to say that 'all you need is two clients, one to say yes when the other says no'. Richard Green rated the Hallsborough as the third gallery dealing in old masters after Agnew's and Colnaghi's.

Harold Leger was more typical of the post-war mid-level Bond Street picture dealer, operating on instinct with no formal training: 'I have a tickle in my palm that's a good picture,' he would say. Lillian Browse, who applied for a job at Leger's in the 1930s, called him one of the world's greatest optimists, something that would later get him into trouble with the Keating scandal. Leger told her baldly, 'If you really want to, you may work in the gallery, but of course I shall not pay you a salary.'[2] To Browse, the terms seemed perfectly fair in view of her complete lack of training, and she entered the art trade. The staff at Leger's comprised an elderly manager, a secretary and two porters. At the top of the building was a restorer known as 'the Baron'. The secretary warned Browse not to fall in love with 'Mr Harold', which she took to assume that he was her own private property, a fact of which he was entirely unaware. Instead, she got on very well with 'Mr Harold', finding him jovial, smiling and rather innocent. Like Mr Micawber, he was sure that something would always turn up.

Browse's first six months were spent filing photographs, the training on which half the art world begins. Art books barely existed at that time, and those that did were mostly in German. For Browse, 'the whole art trade had its well-defined boundary in that the sale room catered for dealers, while they in their turn sold to private people and museums, and not infrequently to each other. This was a straightforward and healthy position – everybody knew where they stood.'[3] She described the week as punctuated by sales at Sotheby's and Christie's. Mondays for 'run of the mill' works and Friday for the finer paintings. Sotheby's held their picture sales on Wednesdays, whilst the other London auctioneers, Phillips, Son & Neale, Robinson & Fisher, Bonhams and Fosters, each had to find whatever slot they could.

When David Posnett joined Leger in 1966, he recalled that his first task in the morning was looking at the ads column in *The Times* for forthcoming sales, and ordering catalogues. When these arrived, they would scrutinise the perfunctory cataloguing, and where possible the imperfect black-and-white illustrations, and decide what to inspect. In order not to give their hand away when they identified a possible 'sleeper', they might commission a bid from Mr Betts, an elderly, whiskery gentleman who lived by buying pictures anonymously for dealers.

The London trade loved a good sleeper, especially when it was at an obscure country sale. They always hoped they would be alone in their insight, although this was rarely the case, as the story about a loveable and respected rogue, Monty Bernard, illustrates.* Bernard 'looked and behaved like an old-fashioned bookmaker, and dealt in English pictures, especially eighteenth-century views of London. But he was a kindly soul and generous to young men who came in search of information, photographs, or pictures to sell. "Well, my boy," Monty would say, with his thumbs in his waistcoat "what can I do for you?"'[4]

Sometime in the late 1950s, there was a house sale at Chew Magna in Somerset. Rumour had spread around the trade that there was a sleeper in the sale, a potential Corot. To Monty's dismay, the train from London was packed with colleagues who settled down to a game of poker in the restaurant car. Changing trains at Bristol to a branch line, Monty bribed the engine driver not to stop at Chew Magna. Nobody noticed Monty's absence until they flashed past their station, pulling the communication cord to no avail. 'Where's Monty?' they cried. When they reached Weston-super-Mare, which was some way off, they were surprised that there were no taxis, which had all been ordered to go on a long journey by somebody on the telephone from Chew Magna. By the time they got to the sale, Monty had purchased the picture and disappeared.[5]

The most spectacular sleeper merchant was undoubtedly Julius Weitzner, who burst onto the London scene in the 1950s selling mostly Italian baroque paintings. It was rumoured that

* Monty Bernard was a Brighton runner who eventually opened his own gallery in Ryder Street.

he had left New York because of tax problems, but according to one dealer, 'The New York dealers didn't like him, whereas the London dealers, although they often raised their eyebrows, admired and accepted him.'[6] Weitzner controlled both the London auction ring and the Italian art trade in London. The Italians in London had a reputation for cash deals and illegally exporting paintings from their homeland. As Julian Agnew put it, 'The Italians regarded their export laws as being fascist laws, and therefore they did not have to pay any attention to them.'[7]

Weitzner divided people: those who knew him loved him, but to those who didn't, he was the sinister figure at the centre of the ring. Born in Boston in 1896, he had studied chemistry and played the violin in a local orchestra, where he met his wife, Ruth Klug, a concert pianist. His first job was as a salesman for pigments, which brought him in touch with artists and restorers. He started his career in the art world by finding an unrecognised Goya *Self-Portrait* in New York before the war. He enjoyed cleaning his pictures himself and this gave him an advantage. Looking nonchalant during sales as he smoked his pipe, he would buy a group of pictures in a sale in collaboration with a posse of dealers (usually Italian). Having them delivered to his house, he spent the afternoon cleaning them, so that by the time his partners came to discuss who would take which picture, Weitzner knew them intimately.

Weitzner sold pictures from his house in Farm Street, which, apart from being a warehouse, became something of a cultural salon. One visitor remembered that 'paintings, sometimes piled ten-deep against the walls, were presented, pell-mell, many unattributed and thus the subject of passionate controversy and discussion'.[8] He and his wife kept open house every evening between 5 and 7 p.m. Here you would meet American museum directors, collectors and other dealers, but never grandees like Geoffrey Agnew, who were his mortal enemies. To Agnew's, Weitzner was 'the crook of crooks', and they would not deal with him.[9] He wrote his invoices on the back of an envelope to avoid the taxman but always kept a carbon copy.[10]

To the dealer Alfred Cohen, 'Julius Weitzner bought all the interesting pictures, even more than Agnew's.' Cohen remembered bidding at the horseshoe table when he got a tap on the shoulder from Weitzner, 'who only needed to wink to make clear that we were going to be buying the painting in shares. Although I had bought the picture, he rang the next day saying that he had taken it away and would be in touch. Two years later, he rang to say that it was sold and for a very good price.'[11] Weitzner's methods were often gangsterish, and he once turned round to the young dealer, Jack Baer, during a sale with the menacing words: 'Stop bidding boy or I'll break you.'[12] He would sometimes soften his threat with 'I just didn't want you to buy a lot of rubbish'.[13]

His most famous 'Lord of the Ring' story was at the house sale of Aldwick Court, Somerset, organised by a local firm, Bruton Knowles, in 1968. A little *Madonna and Child*, possibly by the thirteenth-/fourteenth-century Italian painter Duccio (which would have been a rare find), was catalogued as Sienese School fifteenth century, despite an inscription on the back giving the artist's name. Julius Weitzner bought it in the room for £2,700. Afterwards, there was a knock-out at a local hotel with three other dealers, all members of BADA, where Weitzner paid a rumoured £28,000 before offering it to the Cleveland Museum in Ohio for £150,000. The picture was refused an export licence and went to the National Gallery London, where it is now reattributed to Ugolino di Nerio. The subsequent enquiry was met with a 'wall of silence' from the trade and the Society of London Art Dealers. Both Hugh Leggatt and Geoffrey Agnew were named for withholding information by Lord Glendevon in the House of Lords debate.[14] The art world still hoped it could contain the damage and protect its image.

When in 1971 Titian's *Death of Actaeon* was sold at Christie's by the Earl of Harewood, Paul Getty, who had a preference for bargains, had finally realised he had to pay proper money to buy good pictures. He gave his curator, Burton Fredrecksen, a bid of just over £1 million. Against him was Julius Weitzner in partnership with the German dealer Heinz Kisters, whose prearranged limit was also £1 million. As so often happens on the day, Weitzner was pushed well over his limit and secured the painting for 1.6 million guineas (approximately £1.68 million). The applause was considerable, but Klisters, who had tried to stop Weitzner exceeding their agreed limit, refused to pay his half. In a time-honoured trope, Weitzner told the press that he had 'bought it for his daughter' and wondered what to do next. Nobody believed that Weitzner had a client for the picture. His calculation seems to have been that the underbidder was representing Getty, who realised that he could save money by losing the public auction and doing a private deal. As the bids were going up in 100,000-guinea increments, this was plausible.[15] Fortunately for Weitzner, Getty was annoyed at being outbid and, after the sale, agreed to pay the equivalent of one more bid, so after an anxious few days Weitzner made a profit. To Getty's disappointment, however, the painting was stopped at export and, after a public appeal, entered the National Gallery in London.

Weitzner's most spectacular sleeper was at Bonhams and well demonstrates his cunning and ruthlessness. A painting sent for sale from the boardroom at Pearson's was believed to be a huge copy of the Rubens *Daniel in the Lions' Den*, presumed to be taken from a well-known engraving. It was offered with an estimate of £1,500 and was too large to go into the showroom, so it was parked at the top of the stairs in the corridor. Several dealers had a suspicion that this might be the original, and so Weitzner had to act swiftly. To everyone's dismay, it was announced on the day of the sale that the painting had been withdrawn, presumably, the trade speculated, because it was going to be reoffered fully catalogued later. In fact, Weitzner had rung up Bonhams with an offer of £2,450 if it was immediately withdrawn from sale. His offer was accepted. The sum had been shrewdly chosen, because it was just below the export licence declaration threshold of £2,500. Taking no chances, Weitzner exported the picture within twenty-four hours, even before the sale took place or anybody knew what was going on. As a result of this debacle, the law was changed for export from the *price* to the *value*. To his enemies, Weitzner was an evil genius, and though many in the trade envied his success, some acknowledged that 'He simply had more knowledge and more chutzpah than the rest of us.'[16]

Dutch paintings formed the backbone of the old master market and the largest component of auction sales. There was an abundant supply of them from country houses and they presented fewer export and attributional problems. A talented group of dealers operated in this area, notably Duits, Leonard Koetser (known as Tim) and his son, Brian, Edward Speelman, and in the younger generation, Robert Noortman and Johnny van Haeften. However, to Christopher Wood, 'The king of [Dutch] old master dealers in the 1960s was Eddie Speelman', whose son, Anthony, continues the business today.[17] The firm's great achievement was the formation of the Harold Samuel collection, eventually presented to the Mansion House, where the superb Dutch paintings can be viewed today. Edward Speelman operated from a third floor in Piccadilly, one of the first major dealers not to feel the need for a shop front. As was so typical after the war, he took his batman with him as a porter and factotum. He also had a restorer in the same building.

In 1952, Speelman had met Harold Samuel, the property tycoon who was to make his fortune. The Speelman family were collectors, and they faced a predicament when they were able to buy Ruisdael's *Castle Bentheim*. As the dealer told his son: 'That is a painting I would have loved to have kept for my own collection, but when we have a client like Harold Samuel, that is something you can never do.'[18] He also bought two superb Rembrandts for his client. Samuel was one of those collectors, like Paul Mellon, who never bargained, either taking or rejecting a picture. In 1946, the Speelman logbook records 115 purchases, the following year seventy-three and eighty the year after that. These were as likely to be drawings as paintings. Edward Speelman once commented that, 'A collector is a client, but a client is not necessarily a collector.'[19]

Speelman taught his son Anthony the lesson that what seems outrageously expensive at the time turns out to have been the best buy when you look back several years later. It echoes a saying of David Carritt, that 'the best bargains are those that you pay the most for'. Speelman changed the conduct of auctions forever when, on 2 May 1963, he bought Frans Hals's *The Merry Lute-Player*, over a crackly telephone line from a Parke-Bernet auction in New York. This event prompted the following headline in the *Evening Standard*: '£214,850 DEAL ON THE PHONE: Piccadilly man buys at New York auction'.[20]

The trade was full of eccentrics, one of whom was Mr Dent, who liked to offer presents to those he met in the saleroom. One day, he presented Peter Wilson a toffee apple, which Wilson accepted without having any idea what to do with it. Mr Dent lived in Maida Vale in a house filled with works of art. He began his career as a porter at Christie's and became a scruffy, familiar figure in the salerooms, with a propensity for opening the bidding on major paintings by shouting out 'I'll give you five pounds'. He was almost immediately knocked out, but he did occasionally succeed in buying a few pictures by this means. His highest bid was usually two guineas, and he was said to have a warehouse where he kept all his two-guinea pictures as his pension fund.

The polar opposite of Mr Dent was the refined figure of Count Andrew Ciechanowiecki, always known as 'chicken and whisky', who ran the Heim Gallery in Jermyn Street. A tall man with a quiet manner and a foxy look, Ciechanowiecki was a successful and pioneering art dealer whose main interests were his native Poland, the Catholic Church and the Knights of Malta – in that order. He brought Mittel-Europa culture, stringency and languages to Jermyn Street, and one scholar suggested that, 'To write about the Heim Gallery in London from its opening in 1966 until its final closing in 1995 is to evoke a past recent enough for many readers to remember and at the same time a world as distant to us as New York's Gilded Age or Paris's Fin de Siècle, as described by Henry James or Marcel Proust.'[21]

Andrew arrived in London in 1961 with £2 in his pocket, a list of contacts, unbounded curiosity and considerable charm. Within a year, he had acquired a directorship at the most traditional English furniture shop, Mallett's, in their new showroom at Bourdon House in Davies Street. Here he began to organise original and surprising exhibitions of French nineteenth-century animal sculpture, notably works by Jules Dalou. His wide-ranging expertise was legendary: from Portuguese baroque furniture, Neapolitan paintings and renaissance bronzes to anything else that caught his eye. Michel Laclotte, director of the Louvre, turned to him one day and said, 'You are *also* a dealer!'[22] He eventually gained British nationality, partly thanks to a galaxy of high-born sponsors.

In 1965, Ciechanowiecki left Mallett's to co-found the Heim Gallery at 59 Jermyn Street with François Heim, a leading French dealer in old master paintings. Their first exhibition, *Italian Paintings & Sculptures of the 17th and 18th centuries*, opened in 1966. This set a pattern whereby they would organise two exhibitions a year when Andrew wasn't travelling to meet European and American museum directors. He trained a group of notable young scholars, including Bill Rieder,

Alastair Laing and Alex Kader. Heim's opening parties were a sophisticated mixture of royalty and scholars; Ciechanowiecki had a weakness for exiled royalty and accepting medals. Their exhibitions would be reviewed in *The Burlington Magazine* and indeed some of the entries were written by eminent scholars like Sir Denis Mahon. His catalogues were exemplary, bringing museum-standard descriptions to the art trade (Colnaghi assumed that educated people knew enough) and they vastly extended the range and taste of what people saw in London, as when they put on a show of Thorvaldsen drawings. According to one scholar, 'the summer exhibition in 1971 modestly called *Fourteen Important Neapolitan Paintings*, was, without question, one of the most important exhibitions of old master paintings mounted by a commercial gallery in the second half of the twentieth century'.[23]

Ciechanowiecki was obsessively discreet and avoided buying at auction. The great repositories of baroque paintings in England tended to look towards Colnaghi and Agnew's, so he would search the Continent instead. This led to his greatest humiliation, the sale of four Rubens tapestry cartoons to the National Museum of Wales for £1.2 million in 1979, with an alleged Swiss provenance (which suggested to some an illegal export). The attribution was soon rejected, and it was revealed that images of the cartoons had been published in the 1920s hanging in a palazzo in Milan. In 1986, Andrew sold the Heim Gallery but went on dealing in a small way; however, by 1995, a stroke forced him to give up completely. It is estimated that in its thirty-year life the Heim Gallery handled over a thousand baroque paintings, many of which ended up in museums.

If foreigners like Weitzner and Ciechanowiecki brought knowledge and energy to a London trade still obsessed with the country house world, David Carritt remained the most brilliant native-born operative in the old master sphere. When he left Christie's in 1970, the Belgian banker Baron Leon Lambert founded Artemis as a vehicle to harness Carritt's great gifts. Artemis was an art investment company originally formed by Lambert and the Parisian collector Baron Elie de Rothschild. The idea was simple: to buy works of high quality, lend them to museums where there would be few expenses or insurance costs and eventually sell them for a profit. This followed the pattern of the British Rail Pension Fund, but the business model soon had to be abandoned, because art dealers need turnover to pay the bills.

The range of Artemis was impressive, with old master paintings and drawings, Benin heads, Rembrandt etchings and Roman antiquities. In April 1985, they bought the Northampton Mantegna at Christie's on behalf of the Getty for £8.1 million, a record price for a painting at auction. They preferred to make their purchases directly from private owners, and to that end they had the most impressive international board of directors and associates of any London art gallery. These included the New York dealer Eugene Thaw, the antiquities collector George Ortiz, the Parisian man of taste Alexis de Redé and the international dealer and collector Heinz Berggruen. The urbane Tim Bathurst was lured away from Arthur Tooth to be managing director at Thaw's suggestion. David Carritt was not really interested in selling, so Thaw, who had unique access to the American market, found most of the buyers himself.

Artemis was involved in a murky antiquities story concerning a bronze statue of a youth which came via a Munich-based antiquities dealer, Heinz Herzer. In 1964, the bronze statue had been hauled up by fishermen, supposedly in international waters just off the Italian seaport of Fano on the east coast. It was of an athlete and attributed by some to Lysippus. The 'Fano Athlete' was passed from a local carpet dealer to a furniture restorer, who paid nearly $4,000. Artemis paid a reported $700,000 and sold it to the Getty in 1972 for a rumoured $1 million. Harry Ward

Bailey, the ex-Christie's representative in Rome, ran the antiquities department at Artemis and was said to have lived on his profit thereafter. Artemis never fulfilled its promise, however, and when David Carritt died in 1982, the firm lost its glamour and changed hands.

Without question, the great survivor of the period is Richard Green, who spans two chapters of this book in his longevity. As Christopher Wood put it, 'His career has certainly been the most remarkable of any dealer I have ever known.' Green's turnover and buying power made him the engine of the London art market, and he instinctively knew what it wanted. The barometer of taste and the market for over half a century, he has always seen over the horizon to where things were going and ridden on the crest of that wave. His stock used to be distinctive for its high-gloss varnish and top-of-the-market prices, which his clients never seemed to mind paying. Green ploughed a separate furrow, or, as one competitor put it, 'He was the grand dealer for people who are not grand but have money.'[24] The success of the gallery was in making art accessible to people not used to buying it and who were nervous of entering the portals of Agnew's. Midlands manufacturers and Norfolk farmers would come to London and see a large selection of attractive, well-presented paintings. The brand became trustworthy and a promise of quality.

A conversational minimalist who has on occasion been known to be a little overbearing, Green has been admired, even fondly, by the London trade, who recognise him as king of the market. As one dealer said of Richard, 'you were never going to get the better of him but he could equally be charming albeit in a rather gruff way'.[25] Richard started at the age of fifteen with his father James Green, affectionately known in the trade as 'Slasher Green', who sold sporting art, nineteenth-century still lifes and genre paintings of cardinals. James had two sons, Richard and John, who joined the Life Guards and then went into the property market, eventually ending up running one of his brother's galleries. Richard's first gallery opened in the early 1960s in Dover Street, where he followed his father's tastes.

During the 1970s, people primarily associated Richard Green with sporting art, and he would usually advertise a Herring, a Ferneley or occasionally a Nasmyth landscape in *Country Life*. He had much competition in this field: Spink, Ackerman, Sidney Sabin, Peter Johnson and Frost and Reed were all selling sporting pictures and, like Green, held exhibitions. Perhaps Green squeezed them out. He certainly attracted their clients, because he had more paintings for sale, and when this market began to dwindle with changing tastes, he moved more easily into new areas.

In addition to selling old masters, English sporting and marine paintings, Green bought Victorian pictures, particularly moonlit scenes by Atkinson Grimshaw, and later he partly shifted into twentieth-century art. In his own words, 'British sporting art has always been a section, but certainly not the whole of the business. Nasmyths and Leader – a few, but not particularly. We have concentrated on old masters, Dutch flower pieces, still life and genre; Italian *vedute*, nineteenth-century European, Scottish Colourists, then the rise of Modern British market from the 1990s onwards.'[26] Preferring attractive and decorative subject matter, he has eschewed difficult, academic or overtly religious iconography. Green has made very few mistakes of direction and usually got his timing right: 'I take risks only on tried and trusted artists,' he has said, 'and just stick to what I understand.'[27]

A believer in turnover, Green preferred to do a deal and move on. Part of his success depends on buying reliable names and shirking attributional problems, favouring resaleable pictures with no question marks over their condition. Green believes that if you buy pictures at the top of the market, there will always be a customer looking for good quality, good condition, attractive subject matter and good provenance. His ingredient for success is 'buy the best pictures available and have a strong presentation, research and selling team, which enables us to place them with collectors'.[28] When I enquired how many pictures the gallery sold each decade, the answer was an astonishing record: 1960s – approximately 3,000; 1970s – 11,700; 1980s – 6,500; 1990s – 4,800. The later lower figures reflect the move to selling fewer paintings at higher prices.

Green has always been one of the best-financed dealers. When most dealers were short of money, Green still had serious capital to spend, and could dominate the market. Backed at various times by Jim Slater (who became a client and then an investor) and Lord Swaythling, above all, Green had a huge credit arrangement at Sotheby's and no doubt at Christie's too. He negotiated the longest credit facility in the trade, and often sold paintings before he had even paid for them. Sotheby's gave him so much credit that they felt it necessary to insure his life.[29]

Richard Green occasionally sold to museums, for instance Meindert Hobbema's *A wooded landscape with travelers on a path through a hamlet* to the Getty Museum, and J. M. W. Turner's *Sheerness* to Houston. He also sold to the British Rail Pension Fund, whose buyer, Anna Maria Edelstein, described seeing pictures 'while he was breathing heavily behind me. He was not pushy, was very straight and had good pictures.'[30] She bought a pair of Frans Post views of Brazil from him. Adapting continuously, Green prospered where others failed.

Beneath these successful dealerships, the art trade attracted scores of young recruits. During the 1960s, becoming an art dealer was seen as an alternative lifestyle, a bit Bohemian, independent, a glamorous alternative to the corporate ladder. This was especially true for contemporary art, but even in traditional areas the younger generation saw themselves as disrupters to the Agnew's establishment: William Drummond, Neil Hobhouse and Andrew Edmunds all thrived in this freewheeling world.* A generation of well-educated dealers, they could be found delving through the photographic box files of the Witt Library at the Courtauld Institute, and no longer relied on anything like Harold Leger's proverbial 'tickle in my palm'.

* In 1986, Sarah Ferguson, the future Duchess of York, went to work for William Drummond, who found her 'unbelievably efficient, helpful, outgoing and thoughtful'.

CHAPTER 6

Modern Art: Mostly Cork Street

'The art trade in those days was small and exclusive; it had not yet become a fashionable profession or a field for the private speculator.'[1]

LILLIAN BROWSE

Home to only a handful of galleries in 1945, by the mid-1970s Cork Street had as many as twenty shops and would become as synonymous with modern art as Bond Street was with old masters. It was a cosy world of gallery owners, artists and collectors who socialised within a close fraternity, even after the Marlborough Gallery came and upset the applecart. Henry Roland compared Cork Street to the rue de Seine in Paris, and indeed it had a Francophile character that was slow to wake up to the new American art movements. Apart from the School of Paris, art galleries were focused mostly on native-born artists whose reputations had been established before the war, including Paul Nash, Henry Moore, Ben Nicholson, John Piper and Graham Sutherland. In 1945, Francis Bacon made his grand entrance onto the scene with *Three Studies for Figures at the Base of a Crucifixion* (1944), exhibited at the Lefevre Gallery (now in the Tate). Over the next decade, he became the darling of the modern market.

The dealer who held sway on Cork Street before the war

was the much admired and loved Oliver Brown at the Leicester Galleries. As Lillian Browse observed, 'Few people today can know at first hand the significant role that the Leicester Galleries played during the period between the world wars … Oliver Brown and Cecil Phillips were the first London dealers to stage exhibitions by the great French masters of the 19th and 20th centuries … besides taking chances on a handful of sculptors such as Moore, Epstein and Hepworth.'[2] After handling the Hugh Walpole estate in 1945, however, the Leicester was not to become a major post-war player. By this time, new dealerships and institutions were being formed to promote contemporary art. The ICA opened in 1948 and moved into its own premises in Dover Street in 1950; however, under the presidency of Herbert Read and the chairmanship of Roland Penrose, its interest in surrealism and cubism appeared a little outdated to the younger generation. As director of the Whitechapel Art Gallery, Bryan Robertson put on exhibitions that were very influential, showing all that was new from Europe and promoting a more transatlantic outlook. A well-educated German-speaking émigré community congregated around the St George's Gallery, a bookshop and gallery that acted as a social hub and gave employment to two future art market stars: Erica Brausen and Harry Fischer. The gallery displayed German expressionists and artists such as Siegfried Alva, Heinz Henghes and Oskar Kokoschka.

It was partly because so many post-war dealers had come from the Continent that Cork Street would initially have a European emphasis. Typical were Charles and Peter Gimpel, brothers of Alsatian descent whose family held one of the great international dealerships before the war. Establishing themselves in London in 1946 as 'Gimpel Fils', they used their father's surviving stock for their first exhibition: 'Five Centuries of French Painting'. The Gimpels brought with them the concept of the artist-dealer relationship pioneered by Paul Durand-Ruel, the great Paris dealer who had collaborated with the impressionists. In return for an annual allowance to an artist, the gallery would expect an exclusivity. Working on this model, Gimpel quickly attracted an important stable of artists, including Lynn Chadwick, Kenneth Armitage, Alan Davie, Barbara Hepworth, Ben Nicholson and Victor Pasmore. In time, they were all to be poached by Marlborough, except Davie, who stayed with Gimpel.[3] The brothers became disillusioned when they put on an exhibition of Abstraction which sold nothing, and they asserted that the British public were forty years behind the times.[4]

If Gimpel had the stock to carry them through, the legendary Erica Brausen 'had to rely on flair, and that she held in trumps'.[5] The dealer John Kasmin commented that, in those days, modern art dealers were 'gay, Jewish, or menstrual women – Erica was all three!'[6] She left Germany in 1929 for Paris and then Mallorca, where she ran a bohemian bar, before pitching up broke in London just before the war. In London, Brausen began by serving an apprenticeship at the St George's Gallery and the Redfern. The Barbican exhibition catalogue, *New Art in Britain* (2022), neatly summarises her story: 'In 1947 she established the highly influential Hanover Gallery, which was at the centre of Soho's queer art network: Brausen was openly lesbian and her partner Catharina "Toto" Koopman, a survivor of Nazi incarceration, helped her run the gallery, while financial backing was provided by the wealthy gay Anglo-American, Arthur Jeffress.'[7] Dynamic, charming and a good linguist, Brausen would give tea parties in French, calling everybody 'darling'. Although kind-hearted, she had a sharp tongue and could be rude to young artists, who, having put out their art for her approval, she would ask: 'where are ze pictures? I don't see any pictures.'[8]

According to her gallery assistant, Yvonne Robinson, Brausen was a cosmopolitan who never really understood the British. She lived a comfortable life between a house in Greece on the Ionian Sea, a flat in Eaton Place and another in the seventh arrondissement in Paris. She once said that she was an art dealer because 'I like to make money and have fun'.[9] Although the Hanover Gallery opened with an exhibition of Graham Sutherland, Brausen also had an arrangement with the Paris dealer Iolas, who supplied her with French artists like Maillol and introduced her to Giacometti. With its thick carpets, Regency desk and luxurious framing, the Hanover Gallery was the antithesis of the Hampstead-like austerity of the London Gallery in Brook Street, where George Melly worked before becoming a Jazz musician.[10] According to Melly, Erica's shrewdness lay in realising that contemporary art was about to become fashionable on an international scale, and must be presented to advertise not only the collector's taste, but also their money.

As Melly put it, Brausen's masterstroke was 'to grab Francis Bacon when most people recoiled from him in horror, even though it transpired eventually that she was only fattening him up for the Marlborough Gallery to gobble up later'.[11] It may have been her sale of *Painting* (1946) to Alfred Barr at MOMA that won Bacon's loyalty, and Erica represented him for a decade from 1948, staging his first solo show in 1949. Robert and Lisa Sainsbury, her main clients, scooped up much of the artist's best work. Brausen was also an early exhibitor of Lucian Freud and showed his work in 1949, although she found him tricky and he annoyed her by selling two paintings ahead of the gallery's 1952 exhibition to John Rothenstein at the Tate. Other British artists she showed were Eduardo Paolozzi, William Scott, Reg Butler and Richard Hamilton. Erica was heartbroken when Bacon left her without explanation at the end of 1958, and moved to the Marlborough Gallery.[12] Although Freud claimed that he had persuaded Bacon to leave, it is equally likely that the separation came when she had refused to pay his gambling debts. Whatever the reason, Yvonne Robinson recalled that, thereafter, Erica referred to Marlborough as the Gestapo.

Lillian Browse was a rare youthful British woman dealer when most female dealers in London were middle-aged Germans.* After a trade apprenticeship at Leger, her career changed thanks to a chance telephone call in 1945 from Henry Roland (formerly Rosenbaum). Roland announced that he and Gustav Delbanco wanted to start a gallery in the West End, selling both old and modern masters, and offered her a partnership to run the modern side. Securing premises with a cheap lease at 19 Cork Street during the doodlebug era, when few invested in London property, Roland, Browse and Delbanco was born.

As a gallery, the firm remained small, because they disliked overdrafts. Fairly catholic in their tastes, they sold paintings by Courbet, Lovis Corinth and Fuseli; sculptures by Rodin and Emilio Greco; and drawings by Ingres and Constantin Guys. William Nicholson and Walter Sickert – on both of whom Browse had written monographs – were her favourite artists. Dressing in an elaborate Edwardian manner, she mildly annoyed Roland, who said that, 'She had tremendous will-power, and I always had the feeling that she made people buy what she wanted them to buy.'[13] Of his other partner, Gustav Delbanco, Roland observed that he had verbal diarrhoea: 'He talks and talks and talks and never stops.'[14] The result was that the partners made it a rule to very rarely meet socially, preventing jealousy among their spouses or between each other. Among their clients was Henry Moore, who bought three Courbets, a Rodin and a Redon drawing. Their proudest exhibition during the 1950s was of the 'Fauves', still rare in England, and they sold the Derain of the *Port of London* to the Tate. Browse was later to donate her own collection to the Courtauld Gallery. The only time a queue ever formed outside

* Lillian Browse had a distinguished war, organising exhibitions of modern artists at the National Gallery.

the gallery was for an exhibition of sixty watercolours by Paul Klee, on loan from the artist's son. Roland, Browse and Delbanco only charged 33 per cent to contemporary artists, the same as the Fine Art Society, where other galleries charged 40 or 50 per cent.*

If you wanted to buy an expensive impressionist painting in post-war London, you would visit either Dudley Tooth or the Lefevre Gallery, next door to each other in Bruton Street. It was said that 'during the 1960s everybody in Eaton Square owned a Bonnard acquired from the dashing and well-connected Dudley Tooth at Arthur Tooth & Sons'. Although he sold impressionists to the tycoons of Belgravia, perhaps his greatest legacy was employing David Gibbs and Peter Cochrane, who were instrumental in bringing American artists to the gallery. A financial notch above Tooth, the Lefevre Gallery numbered the international set and particularly the rich Greeks (sometimes responsible for 40 per cent of their sales during the late 1960s) as their clients. This made Lefevre one of the few galleries to hold their own against the great tidal wave of Marlborough. The firm had an illustrious history – their co-founder Alex Reid had known and been painted by Van Gogh – and held a basement full of masterpieces. Typically, Lefevre would have in stock a Van Gogh, some Pissarros, a Monet or two, a Sisley and a group of Degas bronzes.

Run in those days by Gerald Corcoran, Lefevre was also a major player in the British artist market. Bacon had his first exhibition there in 1945, included in a group show alongside Frances Hodgkins and Graham Sutherland. The gallery also sold Edward Burra, Henry Moore, Barbara Hepworth, Ben Nicholson and L. S. Lowry. In the beginning, Lowry was something of a commercial dud for Lefevre: his first exhibition in

* When her two partners retired, Browse opened a new gallery with William Darby in 1977 with a show by Euan Uglow.

1939 sold nothing, followed by two more unsuccessful wartime exhibitions. However, the gallery persisted, and the artist repaid them by staying with them until his death, an almost unique record.* Lucian Freud also had his debut at the Lefevre in 1944. As the grandson of Sigmund Freud, there was much interest in him and he attended the preview, something he would not do again until very late in his career. Herbert Read, the champion of modernism and abstraction, refused to give a commendation for the exhibition, claiming (inaccurately), 'I only write prefaces for refugees.'[15]

If Lefevre was grand and plutocratic, the Piccadilly Gallery was small and lovable with unusual interests: symbolism (particularly Gustave Moreau), secessionist artists, aesthetic movement, 1890s erotica, art nouveau and the twentieth-century heirs of these movements. The gallery exhibited artists such as Max Beerbohm, Gwen John, Eric Gill and Stanley Spencer. Founded in 1953 by Godfrey and Eve Pilkington with Christabel Briggs in the Piccadilly Arcade, it soon decamped to Cork Street. Godfrey Pilkington's obituary described him as 'short, bald, and compact, a Pickwickian character, amusing and kind'.[16] One morning he found himself the victim of high jinks. His Cork Street neighbour, James Mayor, recalls:

> On my way to work I remember seeing Godfrey Pilkington on all fours trying to get into his gallery. Robert Fraser [the neighbouring dealer] came up to me rather sheepishly to explain that he had been a bit miffed with the Cork Street giant, Leslie Waddington, the night before, and rather the worse for wear, he decided to fix Leslie's door with superglue. Unfortunately, he got the wrong gallery.[17]

* The artist also had a relationship with the Crane Kalman Gallery in Knightsbridge.

The Mayor Gallery – like Waddington – was one of the 'dynasty galleries of Cork Street', founded by Freddie Mayor in 1925. It moved to Cork Street in 1933, when the irascible Douglas Cooper, a famously disruptive figure in the art world, became a partner and financial backer. The gallery specialised in surrealism and is credited with the first London exhibitions of Max Ernst, Juan Gris, Joan Miró, Paul Klee and Alexander Calder. Cooper had been banned from the other gallery specialising in surrealism, the London Gallery in Brook Street, after dissuading a couple from buying a John Craxton drawing there. Undaunted, he showed up and George Melly, who worked at the London Gallery, asked him to leave. At this point the potbellied collector bawled, 'I don't know you and I don't want to know you … you whippersnapper!' Melly grabbed him by the back of his collar and frogmarched him to the door, with Cooper screaming, 'I've been waiting for this moment; I shall sue and I have a witness!' 'Oh no you haven't,' said the onlooker, 'that's not the way you treat sales staff.'[18]

The London Gallery had been founded in 1936 by Lady Norton (known as Peter), whose acquaintance with the bauhaus was novel in London. Roland Penrose bought the gallery from her two years later, transferring it to the direction of the eccentric E. L. T. Mesens, and from then on it was spectacularly uncommercial in its operations. Melly, who claimed of Mesens that 'I went to bed with both him and his wife', wrote a hilarious account of working at the gallery in *Don't Tell Sybil* (1997), which describes the freewheeling bohemian atmosphere of the era. One of the London Gallery's first post-war exhibitions was *The Cubist Spirit in Its Time* (1947), consisting of Picassos from the Penrose Collection accompanied by a catalogue including texts by Apollinaire and André Breton. Melly ruefully pointed out that very few people attended the private view, and that, 'The sparse public who turned up that chilly spring evening were an indication the London Gallery was doomed from the start.'[19]

The only gallery that can almost match the Mayor Gallery's longevity in Cork Street is the Redfern, which was founded in 1923 and moved there in 1936. Directed by a New Zealander, Rex Nan Kivell, partnered by an Australian, Harry Tatlock-Miller, it was where Erica Brausen and Peter Cochrane were to cut their teeth. Nan Kivell sold British artists such as Henry Moore and Barbara Hepworth and some French classics, notably Bonnard, Vuillard and, more unusually, Chaïm Soutine. The gallery's most famous exhibition, held in 1957, was curated by the art historian and critic Denys Sutton, with the rather cumbersome title *Metavisual Tachiste Abstract Art: Painting in England Today*. It was a compendium of native artists that included Roger Hilton, Gillian Ayres, Peter Lanyon, Terry Frost, Ralph Rumney and Patrick Heron (who chose many of the works).

The family who came to dominate Cork Street were the Waddingtons. Victor Waddington returned to London from Dublin, where he also had a gallery, and opened a space on Cork Street in 1957. Accompanying him was his twenty-four-year-old son Leslie, who would open his own space a decade later, and by the late 1980s, Cork Street would be home to five Waddington galleries, the subject of another chapter.

Chapter 7

Cuckoo in the Nest: The Marlborough Gallery

'Then in the 1950s came the spectacular advance of Marlborough Fine Art, sweeping all before it.'[1]

John Russell Taylor

No gallery in London was more powerful, influential or disruptive than the Marlborough Gallery in its heyday. For over twenty years, it created and dominated the template for the modern gallery system. For anybody wishing to buy or see modern art in London during the 1950s and 1960s, the Marlborough was the place to go. In 1969, a critic remarked on its power, claiming that 'England had one-and-a-half museums of modern art – the half being the Tate Gallery'.[2] Seen in Hollywood terms, Cork Street dealerships were independents, warm and small, while the Marlborough was the triumph of studio, international and focused on marketing and profits. A whale among sprats, it was regarded with suspicion by its rivals, owing to its predatory instincts and ruthlessly business-orientated attitude to selling art. Marlborough contrasted with the gentlemanly lunchtime approach of other London dealers of the period – and artists flocked to it.

The gallery was founded by Frank Lloyd and Harry Fischer, Austrian Jews who fled their homeland after the *Anschluss* in

1938. Lloyd's family had had an antiques business in Vienna and Fischer had been a bookdealer in the same city. After their internment on the Isle of Man, they met in the Pioneer Corps. It was said that they had bought doughnuts from the village during the day and sold them in the camp in the evening. On Friday nights, they would play cards, and Lloyd always won. Marooned in Britain, Fischer had become depressed, and one day he put his head on a railway line. Lloyd rescued him by reminding him that it was Sunday and there were no trains until Thursday. After the war, Lloyd briefly returned to Vienna, where he owned a petrol station on a corner site in the Russian zone, which he sold profitably to raise money for the gallery.

In 1946, Lloyd and Fischer opened Marlborough Fine Art, which included Fischer's antiquarian book section, at 17–18 Old Bond Street.* One profile described Lloyd as 'a small man with a sharp smile, expressive eyebrows and a knowing manner of speech', while Fischer was seen as 'larger, genial and goggling'.[3] Frank smoked huge cigars and had a twinkle in his eye; Harry was the more cultivated, a large, bumbling figure with protruding front teeth who wrote opera librettos in his spare time. Lucian Freud thought Fischer was 'a shoddy, fat, pompous fool. He came round and borrowed a book of poems and I never got them back.'[4] Despite this, Fischer usually gets better press than Lloyd, who was more entertaining but had an amoral streak. In fact, James Kirkman, who worked for them both for ten years, concluded that 'they were not nice people even if they could be charming'.[5] Although they needed each other, they were evidently not fond of one another: Fischer was better at choosing art, whereas Lloyd had an extraordinary gift for reading people and was the financial genius. 'I collect money, not art,' he said; 'there is only one measure of success in running a gallery:

* Ten years later, this section would become Marlborough Rare Books, and would move to separate premises in nearby Duke Street, St James's.

making money. Any dealer who says it is not, is a hypocrite, or will soon be closing his doors.'[6]

The third in the triumvirate at Marlborough was David Somerset, the Duke of Beaufort's heir, brought in by Lloyd and Fischer to open the country-house doors. Somerset was grand, droll and considered to be the most handsome man in Europe. As one colleague put it, 'he was unbelievably rude to Frank and Harry', and yet somehow, with his charm, he got away with it. Introduced to the gallery through the French painter, Paul Maze, who had the best address book of any artist in London, Somerset brought his fabulous connections. He sold, for instance, Canalettos from the Duke of Buccleuch to Gianni Agnelli, to whom he used to speak at 6.30 a.m. every morning. As Lloyd observed, 'We needed some class, some atmosphere, and David gave it to us.'[7] He raised their game, and Marlborough's openings became very smart, with gossip columnists in attendance and even Queen Elizabeth II showing up on one benefit night.

The gallery that became Marlborough opened in 1946 with £10,000 in capital, and by 1960 it had a turnover well in excess of £1 million and a stock valued at £7.5 million.[8] It was named Marlborough after Lord Ivor Churchill, the aesthete, grandee and one of the first men in Britain to own a Picasso.* The deal that really established the company was selling the Norwegian ship owner Ragnar Moltzau's collection to the Stuttgart Museum. Beating a path to the gallery's door, the international rich also provided backing, whether they were Italian car-makers or Greek ship-owners. There was little trust, however, between the directors and Somerset, who had no idea what Lloyd was up to behind his back, had to sleep with one of the secretaries to find out.

* Surprisingly for a collector who had such pioneering taste, Churchill swapped his collection to acquire paintings by Paul Maze.

A striking aspect of Marlborough's history is the speed at which it built up its stock of modern masters. As Fischer once said, 'the most successful gallery is the gallery that knows where to lay its hands on the rarest pictures at the shortest notice'.[9] In this Marlborough's advantage over nearly all other galleries in London was a familiarity with the private collections of France, Germany and Switzerland.

In 1947, Marlborough presented its first exhibition of French nineteenth- and twentieth-century drawings and paintings, beginning a pattern of seven or eight important exhibitions a year. Soon the gallery was putting on solo exhibitions: Maurice Utrillo, Othon Friesz, Puvis de Chavannes, Renoir, Gris, Monet and an important collection of seventy-three bronzes by Degas. The scale and ambition of the shows was extraordinary. Annual exhibitions of *French Masters of the XIX and XX Centuries* were blockbusters by modern standards. Whilst these artists were familiar to a London audience, Marlborough was also bringing less familiar German, Austrian and other European artists to London. From 1958, they started showing Jawlensky, Kokoschka, Klee, Munch, Kandinsky, Schwitters, Schiele and Nolde. Two groundbreaking shows were the 1959 *Art in Revolt* exhibition of German art between 1905 and 1925 and the 1962 exhibition *Painters of the Bauhaus*. The gallery's first American exhibition, *New York Scene*, held in 1961, included Ellsworth Kelly, Lee Krasner and Helen Frankenthaler. Collaborating with Daniel-Henry Kahnweiler, Picasso's dealer, Marlborough was able to exhibit the artist's recent drawings and bronzes, as well as a retrospective of bronzes by Henri Laurens.

Starting with Géricault in 1952 and Courbet the following year, Marlborough staged major loan exhibitions, with most of the paintings and drawings from the Continent. *Kandinsky: the Road to Abstraction*, held in 1961, was a typical example, showing a mixture of works for sale and loans from museums and the family. Marlborough was adept at nurturing the wives, widows and heirs of artists like Kandinsky, Pollock and Van Gogh, persuading them to make loans. Collectors like Kenneth Clark, who was happy to lend his Seurats, were roped in, and when they needed money they came back to Marlborough to make a sale. This sophisticated business-winning strategy was far ahead of anything being done at Sotheby's and Christie's at the time. Moreover, the exhibitions were arranged and curated to a high scholarly standard.

Without question, Marlborough's most remarkable loan exhibition opened on 16 October 1960: *Van Gogh Self-Portraits* featured fifteen paintings borrowed from the artist's nephew, the engineer V. W. Van Gogh, and the Kröller-Muller Museum. There were queues down the street to see the Van Goghs, and *The Observer* described the opening of the exhibition, commenting on the dichotomy between the black-tie dress on the invitation and the subject matter of a man who never wore a tie. The scholarly catalogue reprinted Kokoschka's 1953 essay on 'Van Gogh's influence on modern painting' and added a reference list with reproductions of the other eighteen self-portraits not included in the exhibition. The gallery formed a good relationship with V. W. Van Gogh but never got a sale from him. They were, however, later to sell the *Self-portrait with Bandaged Ear* from Leigh Block's collection in Chicago.

During the 1950s, Marlborough's main London rivals were Tooth's and the Lefevre, and by the 1960s Ernst Beyeler in Basel. Despite this competition, the gallery still had the pick of the clients: Thyssen, Niarchos, Annenberg, the Koerfer family, the shah's wife Farah Diba, Agnelli and Emile Bührle, the controversial Zurich arms-manufacturer and collector, who bought fifty-two pictures from Marlborough. John Kasmin recalls Agnelli coming to London in 1960 to buy footballers for Juventus, his Turin football club. Although it was a bank holiday, the staff reluctantly all came in, and Agnelli bought an enormous amount. At Marlborough, collectors would be actively wooed to consider a purchase or deliberately denied the chance to acquire a work, depending on the state and standing of their collection. The gallery had a Swiss and Liechtenstein company which was important to sellers; expensive paintings would be exported to Vaduz or Switzerland and then sold back to London, ensuring the profits were made offshore rather than in high-tax Britain. This was legal and common practice at the higher end of the art trade.

Although John Lumley at Christie's recalls that, 'From our point of view Marlborough were mysterious', in some ways they were open – publishing museums they had sold to and artists' estates they had handled, which was novel. By 1963, they had already sold to over forty museums and galleries around the world. They sometimes beat Sotheby's and Christie's at their own game. When Christie's were consulted by the family of the collector Sir Alfred Chester Beatty over Van Gogh's *Patience Escalier*, and they placed it in an auction, it was immediately removed by Marlborough, in cahoots with Lefevre. Although they claimed a prior agreement, it was just as likely to have been the result of a large offer to the vendor, and they sold the painting to Niarchos.

Lloyd appreciated that the supply of great modern masters was limited, and on his doorstep was a talented generation of British artists. The gallery decided to place more emphasis on this area, and in 1958 opened the New London Gallery, for which it is best remembered today. A magnet for attracting living artists, the gallery 'took over the career and management,' as John Russell put it, 'of nearly all the major British artists of international reputation including Henry Moore, Graham Sutherland, Francis Bacon, Ben Nicholson, Barbara Hepworth, Kenneth Armitage and Lynn Chadwick, as well as the profitable but more parochial John Piper'.[10] Looking for a manager for the new gallery, Lloyd interviewed Kasmin in French and liked the fact that he was Jewish and married with a child. He told him,

'If you stay I will make you bigger than Fischer.'[11] Kasmin went to an astrologer before agreeing to join what were even then regarded as the monsters of the art world. Although the astrologer advised against it, he ignored the advice.

For the most part, Marlborough did not cultivate young artists, preferring those who were already internationally successful, like Moore and Sutherland. 'It's a bit like stealing a patent,' said Peter Gimpel, after having groomed and lost Hepworth, Nicholson, Chadwick and Armitage to Marlborough.[12] The gallery nurtured very few talents from the beginning, Kitaj's 1963 exhibition being a notable exception as a first solo show.* In Kasmin's view, they were most keen on John Piper not for his work but because of his mailing list, which was said to be extraordinary. Henry Moore's work made the most money and enabled Marlborough to invest in some new artists, however they did not accept work by living artists on consignment.

In Kasmin's view, their approach differed markedly from the less professional and more patronising approach of many other London gallerists. Artists with whom Marlborough signed contracts could rely on the gallery to promote their work internationally, give them exhibitions around the world and provide them with stipends and cash advances against future works. As he claimed, 'Their systems were wonderful: documentation, photographs, record-keeping, framing, catalogue design, with financial and taxation schemes.'[13] Marlborough was not the first dealership to put artists on salaries. Durand-Ruel and Kahnweiler had done so long before in Paris for the impressionists and their successors. More recently, Gimpel Fils had done it in London, but that didn't prevent their artists from

* Kitaj invented the term 'School of London' for the 1976 exhibition *The Human Clay* at the Hayward Gallery. It was intended to describe a loose group of artists working in London and pursuing figurative painting against the grain of other avant-garde movements of the time.

decamping to Marlborough. Lloyd proclaimed: 'we don't have to go and tempt them; they come to us. We know everything about a painter, about his divorce and his new flats', and in a phrase which would come back to haunt him, 'We are like the old uncle.'[14]

For Martin Summers of Lefevre, 'The magic of Marlborough to the artists was that they had a bigger chequebook, paid artists in advance, as well as having magnificent premises and an international reach of buyers. It was a very slick operation, less interested in giving lunches and having comfortable armchairs than going straight to the business.'[15] Marlborough learned how to make the life of an artist very easy; they collected work from them, they framed it, documented and photographed it, they stored it and prepared exhibitions with museum-like exhibition catalogues and an international sales pitch. The catalogues were beautifully designed in the early days by Guido Morris and later by Gordon House. The first gallery to think in terms of brand marketing rather than schmoozing, Marlborough had no dining room in London to entertain clients. Lloyd and Fischer left that side of the business to David Somerset, who had the added advantage of being a member of the Eagle Club in St Moritz, where he could hang out all day with Agnelli, Baron Thyssen and their ilk.

The catalogue of the 1963 R. B. Kitaj exhibition reveals that the gallery was already the agent for, amongst others, Francis Bacon, Henry Moore, Oskar Kokoschka, Graham Sutherland, Lucian Freud and the estate of Jackson Pollock. Interestingly, Ben Nicholson, who had his first show with Marlborough that year, is not mentioned. Amongst all this modernism, the gallery had regular exhibitions of the traditional painter Edward Seago, whom they took over from Colnaghi. Sutherland complained bitterly about being in the same stable as such a commercial artist. Seago had society contacts including the Royal

Family, and later went to Richard Green. During the 1960s, Marlborough's turnover was 80 per cent modern masters to 20 per cent contemporary, but by the end of the century this had somewhat reversed. Kasmin had been frustrated that he could not persuade Fischer and Lloyd to take on John Latham or David Hockney, whose work Fischer thought scruffy and not up to scratch.

Marlborough's overriding success in London was their relationship with Moore and Bacon. Crucial to this was Valerie Beston, the unsung hero of the gallery (who at some point had an affair with Lloyd). As Queen Bee of Marlborough, Valerie ran the office and was responsible for the welfare of the artists. She held everything and everyone together, especially Francis Bacon. The latter signed a contract in 1958, and although he would not agree to an exclusive, he only sold through Marlborough. Henry Moore equally resisted an exclusive contract, but there was a gentleman's agreement that they would have first choice. Betty Tinsley, the sculptor's devoted secretary, would decide (after discussing with Moore) which gallery would be allowed to sell what. She was a great friend of Valerie, which no doubt bolstered Marlborough's position. Moore was generally looked after by Fischer, and when he left the gallery, Valerie was dispatched to persuade the artist to stay with Marlborough. In future, they had to share his work with Fischer's newly established Fischer Fine Art and others.

Bacon presented special problems to Marlborough. Having settled his gambling debts to encourage him to make the move, the gallery held their first exhibition of his work in 1960. Freud, who made the introduction, was given a painting as a thank you. Valerie adored Bacon and looked after him as best she could. She would have lunch with him every Sunday unless they had had a row, which happened from time to time. Bacon would drop into the gallery once a week to inspect his pictures, beautifully framed by Alfred Hecht, and to see whether they needed any repainting, adjustments or even destruction.* On one occasion, Lloyd was offering one of Bacon's *Popes* to a wavering buyer, who said, "I like it but I'm not sure my wife will and it doesn't really go with the décor in our house." Lloyd replied, "You can change your décor, you can change your wife, but once you own it you won't want to change this Bacon."[16] According to the story the collector bought the Bacon, divorced his wife, and redecorated.

Lucian Freud's first exhibition at Marlborough was held two years earlier than Bacon's, in April 1958. According to David Somerset, Lloyd was not keen on the artist, who was unfaithful and difficult. Freud thought that the partners behaved like spoiled brats: 'They would shout at each other across the street. Traffic would stop. They chucked Eduardo out on the telephone "Mr Paolozzi, if you don't like the way we do business, we don't need you."'[17] The Marlborough was, however, tolerant of Freud's betting habit. Terry Miles, a gallery assistant, remembers the artist taking wads of money from the gallery to a betting shop and losing it minutes later, often sums very much higher than Miles earned in a year.[18] Marlborough preferred reliable artists like Graham Sutherland and Sidney Nolan, and one of their young gallery assistants, James Kirkman, left and became Freud's agent.

The year 1964 saw an initiative heralded by an exhibition entitled *Opening of the Print Department*. To create a new income stream and bring in younger buyers, Marlborough artists were encouraged to take up printmaking. These were generally lithographs, which could be marketed in signed editions, which

* According to Lord Hindlip, Francis Bacon had a clause that allowed him to destroy any work until it went on public exhibition at Marlborough, because it was only when he saw his paintings in the large galleries that he could see their quality.

appealed to those who couldn't afford paintings or sculpture. You could buy, for instance, a Henry Moore graphic for as little as £25. Some artists like Ben Nicholson made etchings, while others adopted screen printing.

Having conquered London contemporary art by the early 1960s, Marlborough began to branch out: Rome, Zurich and, later, Tokyo and Madrid. But most important of all was New York. It was the first contemporary art gallery to become a brand. Lloyd established in New York largely thanks to the lawyer and art collector Ralph Colin. Colin knew everybody and had been executor of the art dealer Otto Gerson. Through Colin, Lloyd acquired the Gerson Gallery from his widow, providing him with a ready-made business in New York.

The introductions to American artists came from a remarkable accountant, Bernard J. Reis, who worked for many of them (often pro bono), including Philip Guston, Willem de Kooning and Mark Rothko.[19] Reis became the accountant of the New York gallery, and this would be the genesis of Rothko signing up with Marlborough. On 10 June 1963, Lloyd threw a dinner at the Waldorf to tell the abstract expressionist artists what he could do for them. The gallery that Lloyd opened that November was on the corner of the Fuller building – one of the smartest and largest gallery spaces in the world. From here he soon had contracts with Clyfford Still, Guston, Motherwell and Pollock, all of whom he liked to visit by helicopter to impress them. Concentrating on the new gallery and only occasionally returning to London (where he would stay at Claridge's), Lloyd would find his nemesis in New York.

When Mark Rothko died in 1970, he left 798 paintings in his estate. His will had been drawn up by his executor, Bernard Reis. The over-advantageous deal negotiated by Lloyd was to pay $1.8 million for 100 paintings, with payments to be spread over thirteen years without interest. The story unravelled over the next eleven years in the New York court cases brought by

Rothko's twenty-year-old daughter Kate, and developed into a sensational scandal. Found guilty of conflicts of interest, Frank Lloyd and the executors of the Rothko estate were hit with fines and damages totalling nearly $9.3 million. Lloyd was banned from doing business in the USA until he had completed a sentence of community work. The gallery attempted to shrug off the scandal, and Anthony Haden-Guest thought Lloyd 'was making an excellent job of acting as though the Rothko case were a minor squall in the corner of some distant landscape'.[20] In the opinion of John Erle-Drax, later London chairman of Marlborough, 'The Rothko scandal had little impact on the London office; we were cocooned from it and we still had the goods.'[21] However, to the rest of the world it seemed that the Rothko scandal had severely damaged the Marlborough brand. As Thomas Gibson put it, 'Marlborough was a battleship that had sunk.' Although many artists left, a surprising number stayed and the gallery continued to serve Bacon, Auerbach, Kitaj, Pasmore and Piper. They sold a stock of German expressionists to pay the fine and the London gallery separated from New York to avoid further reputational damage.

The London gallery was to be embroiled in its own estate problem when Francis Bacon died. Marlborough was alleged by the Bacon estate to have grossly undervalued many of his paintings, reselling them for several times the price they had paid the artist himself. The Bacon case collapsed. Art-world gossip suggested that it was because the gallery knew too much about the artist's erratic affairs, but the case was deemed to have no merit.

Despite the scandal, Marlborough had a remarkable way of holding on to its artists. When Desmond Corcoran tried to steal Kitaj, the artist responded, 'I like being with them they do anything I need.' With sympathetic directors like Tony Reichart, who looked after Pasmore and John Piper, the gallery held on. When Fischer separated from Marlborough in 1971, he was paid off with £1 million. The divorce had been coming for some time and was largely an argument about succession. Fischer took Kossoff and Uglow with him. Marlborough nurtured a distinguished group of future art dealers, including Kasmin, Thomas Gibson, David Bathurst, James Kirkman, Achim Moeller and Richard Salmon. A powerhouse at its zenith, one of the gallery's most abiding legacies was as a training ground for future luminaries of the art world.

Chapter 8

New Directions: The Swinging Sixties

'London was very boring. Suddenly about 1964, it was all happening. An Eruption – a social revolution.'[1]

Robert Fraser

The American actor Dennis Hopper described how, 'That sixties time in London was the greatest. I knew I was in a place where all the creation of the world was happening. The Beatles and the Stones had just happened. They were still playing in clubs, and all these music shows with girls in boxes doing go-go stuff ... the art world, the fashion world they were exploding.'[2] *Time* magazine ran a famous issue on swinging London in 1966. It included a map of the art scene (no longer the art world) featuring the ICA as well as the two leading commercial contemporary galleries: the Kasmin Gallery and Robert Fraser's gallery. These pivotal dealers could attract the Beatles, the Stones, Jimi Hendrix, Marlon Brando and Dennis Hopper to their openings. Although Britain was politically and economically losing its world position, the artist Michael Kidner believed that 'the country seemed to wake up'. London was a major cultural centre, with its museums, art schools, magazines, artists, pop stars and actors. Becoming interchangeable, these worlds created the legend of the Swinging Sixties, of which Kasmin and Old Etonian Fraser were very much at the heart.

The Arts Council, directed by Lilian Somerville, and the ICA, under the directorship of Dorothy Morland, provided an important stimulus to native contemporary art. In 1965, the government announced an increase in spending by the Arts Council for assistance to young artists. The *Sunday Times* and *The Observer*'s new colour supplements ran articles about contemporary artists, their dealers and collectors. Bryan Robertson, John Russell and Lord Snowdon's *Private View* (1965) unveiled this glamorous world. Television began to take an interest too, with Ken Russell's *Pop Goes the Easel.* The critic John Russell Taylor thought the *annus mirabilis* for living artists in London was between 1963 and 1964, when talented artists had only to walk out of art school to be snapped up by somebody.[3] Close to Portobello Market and the painting studios of the Royal College of Art, Notting Hill became the bohemian centre, where artists including Peter Blake, David Hockney and Richard Smith lived. The era was marked by the emergence of a distinct and homegrown London School, which looked to the exciting and disruptive developments across the Atlantic and the new generation of artists in the USA.

By the 1950s, for the first time the USA took the lead in the development of Western pictorial art, and the abstract expressionists indicated the future primacy of contemporary art in the market. The Tate Gallery put on two exhibitions of American art in 1956 and 1959, giving impetus to abstract art in Britain. Bryan Robertson's shows of Jackson Pollock, Mark Rothko and (especially) Robert Rauschenberg at Whitechapel were even more influential. The art scene was waking up to the revolution in American art. The introduction of transatlantic jet passenger services in 1958 not only benefited Sotheby's sales but proved to be very stimulating to contemporary art. It brought American collectors to Britain and encouraged art students like Hockney to go and see for himself what was happening in the USA. There was a developing internationalism among not just London artists but also galleries, who teamed up with their New York counterparts to show works on both sides of the Atlantic. Some, like Marlborough, would establish galleries in both London and New York.

It was two young men at the normally rather conservative Arthur Tooth & Sons who are credited as pioneers in bringing American art to London. David Gibbs acquired a collection of American abstract art and co-managed Jackson Pollock's estate (with Lee Krasner), helping to develop a European audience for the artist's work. His colleague, Peter Cochrane, first saw work by the American artists Sam Francis and Ellsworth Kelly when visiting Paris. Introducing them to London, Cochrane sold their work to the most interesting collector of the period, Ted Power, the Irish radio pioneer who would later become Leslie Waddington's main client.[4] Cochrane was also selling Asger Jorn, Allen Jones and Dubuffet, the French artist who seemed to promise so much to that generation. Leaving Tooth, Cochrane soon moved to Christie's and Gibbs went to the Marlborough Gallery.

No two dealers were to epitomise the Swinging Sixties better than Kasmin and Robert Fraser. The critic David Sylvester summarised the distinctions:

> Kasmin came from the Jewish middle class, went to Magdalen College School, Oxford, had no money of his own. Other key differences from Fraser were that Kasmin was heterosexual and a family man, that he preferred alcohol to drugs, viewed art academically where Fraser had a hip reliance on intuition. As dealers, one of their key differences was that Fraser was master in his own house whereas Kasmin had a backer, the Marquess of Dufferin and Ava, a young Guinness heir.[5]

Kasmin had apprenticed at Victor Musgrave's pioneering Gallery One in Soho. Musgrave was the first dealer in London to show Yves Klein and the Fluxus group and later gave Bridget Riley her first exhibition. Unfortunately, the gallery was never properly financed. Kasmin enjoyed his time there but found himself the 'sex slave' of the owner's wife, Ida Kar, an experience he recalled as 'quite taxing', before he went to Marlborough. One of the defining moments of his life was seeing the Morris Louis exhibition at the ICA in 1960, a show which made him realise that America was the new creative engine. Making his first visit to New York two years later, he met all the key people in that scene.

Kasmin really learnt his trade as the first director of the Marlborough's New London Gallery, which opened in 1960 as the firm's vehicle for showing contemporary art. In Kasmin's view, Marlborough was the first truly professional gallery, but after his frustration at not being able to sign up Hockney, he decided to go independent. Kasmin's best client at the gallery was the young Sheridan Dufferin, who offered to put up £25,000 to start an art dealership in which they would be equal partners, using Kasmin's name. They started in the latter's house in Fulham in 1961 and moved two years later to 118 New Bond Street.

Kasmin was the first contemporary dealer to think seriously about space and the aesthetics of showing modern art. His gallery, a white cube designed by Richard Burton of ABK, had white walls, a high, sloping ceiling and natural top lighting to avoid 'those horrid little "spots" here and there'.[6] The lighting was the most sophisticated in London, with warm and cold fluorescents installed by Anthony Juer to balance the daylight. The Pirelli flooring was equally novel, a type of linoleum first used at Rome airport. To Kasmin, the cool and airy space was 'A machine for looking at pictures in', and some thought it the most beautiful room in London. Not every artist appreciated it, however – Anthony Caro insisted Kasmin put in coconut matting, which caused him to exclaim, 'Unbelievable expense. No wonder I went broke! Just to keep Tony happy.'[7]

One feature of the gallery was the scale of the walls designed for huge pictures – museum rather than drawing room works. Kasmin told the architect, 'I want to show artists no one else wants to here because they're either too difficult, too enormous in picture scale, too expensive (some of the Americans) but are still doing great things; and I want to make new collectors of the young rich.'[8] In the latter, he would be sorely disappointed.

The Kasmin Gallery opened in 1963 with Kenneth Noland's concentric circle paintings. The haunt of pop stars, it became a part of the geography of swinging London when Mary Quant did a fashion shoot there. The gallery acquired a reputation for showing American colour field and new wave British artists. The Americans whom Kasmin exhibited were Newman, Reinhardt, Stella, Noland, Louis, Frankenthaler, Olitski and Poons. With the British artists Bernard Cohen, Robyn Denny, Anthony Caro, David Hockney and Richard Smith, he followed the Marlborough playbook by giving them salaries. Kasmin would stage as many as nine one-man shows a year. Among his favourite exhibitions were Morris Louis's *Stripe* paintings and the 1963 exhibition of Richard Smith's shaped canvases. He was proudest of showing Frank Stella and Anthony Caro. Large and heavy, Caro's new work was problematic to exhibit, and when both Gimpel and Marlborough rejected it, Kasmin took it on.

Associated with discovering Hockney, Kasmin's representation of the artist was uncharacteristic of his taste, which ran more to abstraction. He bought Hockney's *Doll Boy* from the *Young Contemporaries* exhibition in 1961. Intrigued by the painting, he asked to meet the artist. There followed a tense encounter between Kasmin and his wife at home and the artist, still a shy, impoverished, dark-haired student at the Royal College of Art. The art school was equivalent to Goldsmiths in the next generation, Hockney's fellow students there included Derek Boshier, Patrick Caulfield, Allen Jones and the American R. B. Kitaj. Graduating the following year, Hockney had his first exhibition at Kasmin in December 1963. *Paintings with People In* consisted of ten works priced between £250 and £400. The show was a critically acclaimed sell-out. Someone complained that the nudes' breasts were nowhere near where they should be and asked Hockney if he'd ever seen a naked woman? 'A dorn't knogh ars ah harve!' came the response.[9] With Hockney's droll Yorkshire manner and recently acquired zany looks, the exhibition elevated the painter to pop star status. His second show at the gallery, *Pictures with Frames and Still-Life Pictures*, opened in December 1965 and contained some of his famous swimming pool paintings, which perceptive critics recognised as the best works on show.

Kasmin had his zenith in 1966, when he represented four of the five British artists at the Venice Biennale: Caro, Denny, Smith and Bernard Cohen, selected by Herbert Read and Alan Bowness. Despite this success, he was depressed about dealing and the state of the economy at a point when several galleries were on the verge of closing. The respected critic Edward Lucie-Smith recalled that, 'More than one gallery director said to me that it hardly seems worthwhile to spend so much money and energy on running a free show for the British. Sales are made abroad, foreign clients shopping in London.'[10] Kasmin complained bitterly that, 'The enormous rise in the cost of running premises like mine has been accompanied by a severe decline in English collecting recently.'[11] There were never enough English clients. Kasmin thought the problem was that there were not enough Jewish people in England compared to New York. He and Leslie Waddington used to joke about breaking through the WASP barrier.

Kasmin was saved by injections of capital from Dufferin, but when his patron married a painter, Lindy Guinness, who preferred the Bloomsbury Group, Dufferin lost interest in the gallery. His ownership went from 50 per cent to 20 per cent, although he was happy to guarantee at the bank anything by Hockney. Kasmin closed the gallery at the end of 1972 but remained active, continuing to represent important artists, and had three different shops around Bond Street over the next twenty years. Christopher Gibbs took over the old gallery, entirely changing the aesthetic with sixteenth-century rugs and eclectic pieces of furniture.

The dealer with whom Kasmin is always compared is Robert Fraser, and they did run the two most happening galleries in London at the time. Julie Christie and Terence Stamp would appear with Mick Jagger at Fraser's openings, and, as Pauline Fordham put it, by comparison 'All the others were staid, old, traditional, no life, no energy shows. The openings were all bow ties, straight cocktails. Robert was the first one to waken things up … film stars who were in town dropped in. Everybody came who wanted to be part of the scene.'[12] They were friendly rivals; Kasmin describes the difference between himself and Fraser: 'Robert was always very hip, I was a bit square: I was married, I had kids, I didn't take heroin.'[13] Despite this account, a lot of marijuana was smoked at both galleries' openings, and both were accompanied by rock music.

Robert Fraser was an *enfant terrible* who might easily have been the antihero of a novel. Later dubbed 'Groovy Bob' (a term he would have hated), Fraser was a first-generation Etonian whose grandfather had been Gordon Selfridge's butler. His father was a successful financier and a trustee of the Tate. To the waspish John Richardson, Fraser always wore 'an impeccable blazer, very Old Etonian, consciously so. I suspect he had slight chips on his shoulder about his father. There was a streak of the old-fashioned servant in a way … You could see that he was a butler's grandson. That's what one liked about him, this flash side, which he never tried to gentrify at all.'[14] With his charm and pop music friends, he seemed to represent the transition between two eras in Britain. Described as looking like a very hip ex-Guards officer, he became the subject of many artists' works, but he remains a rather enigmatic figure.

After a period spent working in galleries in the United States, in 1962–3 he established the Robert Fraser Gallery at 69 Duke Street, Grosvenor Square, with backing from his father. As Jim Dine described, Fraser 'was like a gentleman dealer, but not really. He was hardly a gentleman and not much of a dealer.'[15] Others disagreed: Richard Hamilton thought 'Robert's was the best gallery I knew in London', and Ellsworth Kelly claimed that 'he was a very courageous and flamboyant dealer'. The difficulty for artists was getting paid, despite the fact that Fraser had a white Rolls-Royce and a chauffeur waiting outside the gallery. He kept going thanks to parental subsidies and by not paying his debts.

Fraser's first exhibition was of Jean Dubuffet, at a time when the French artist was seen as the heir to Picasso and the great tradition. As Bryan Robertson of the Whitechapel Gallery put it, 'The early shows that he put on were simply marvellous because they covered an area that the English simply didn't know about.'[16] Although Fraser went on exhibiting Europeans, including Magritte, Michaux, Bellmer and Klapheck, again it was the Americans and above all the pop artists that would give the gallery its distinction: Rauschenberg, Twombly, Oldenburg, Warhol, Dine, Chamberlain, Ruscha, Lindner and Matta. Perhaps his most important exhibition was *American Scene*, a 1967 show of pop artists which included Andy Warhol, Claes Oldenburg, Jim Dine and Roy Lichtenstein. Dennis Hopper said that 'being in England when Robert had the LA show was the most exciting time I'd ever seen'.[17]

Fraser's stable of British artists was impressive: Richard Hamilton, Eduardo Paolozzi, Peter Blake, Harold Cohen, Colin Self, Bridget Riley and Patrick Caulfield. They appreciated the trouble he took for them even if he was late in paying. When an exhibition of Bridget Riley's drawings looked pallid, he had the whole gallery painted black overnight and the next morning they shone. One of Fraser's greatest coups was to persuade Richard Hamilton to give up teaching and become a full-time artist, but as Hamilton ruefully recalled, 'I got a cheque that bounced.'[18]

With Mick Jagger (who thought Fraser was a tastemaker and a hustler), the Guinness heir Tara Browne, John Paul Getty Jr and Christopher Gibbs, Fraser was at the centre of a hip world. He persuaded Paul McCartney to buy a Magritte of a green apple, which became the basis of the Apple logo. When McCartney asked Robert's advice on the record sleeve for *Sgt. Pepper*, he suggested Peter Blake and Jann Haworth, thereby acting as midwife to the most famous record sleeve of all time.

The most notorious episodes in Fraser's life were his confrontations with the police. The first was when he staged an allegedly obscene exhibition of works on paper by Jim Dine in 1966. The case was preposterously brought under the Vagrancy Act of 1838, an obsolete law which had been designed to punish beggars who publicly exposed wounds and sores, although it also covered the exhibition of obscene images. Bryan Robertson spoke for the defence, but the gallery was fined twenty guineas and ordered to pay the same again in costs.

Fraser's second brush with the law was more serious. As a friend of the Rolling Stones, he was at Keith Richards's house in the country when the police made their famous raid leading to the arrest of Mick Jagger, Richards and Fraser. The arrest inspired Richard Hamilton's *Swingeing London 67*, a seminal work based on a press photograph showing Jagger and Fraser handcuffed together. Robert was sent to jail for six months. Nineteen loyal artists, including Kitaj, Hamilton, Hockney, Allen Jones, Riley and William Turnbull, put on an exhibition as a gesture against the severity of his sentence. After his release, Fraser's heroin addiction gradually took control. As Christopher Gibbs said, 'Robert was far too enthralled with his lower nature. I really do think he was seriously indulgent to his appetites.' One last hurrah before the gallery closed was to give the unknown Gilbert & George an exhibition in 1969. They wanted to be with Fraser, as they found Marlborough too stuffy.

After a hiatus, Robert Fraser reopened his gallery in Cork Street in 1983, but by that time he was already seen as the past. Ill and disillusioned with art dealing, Fraser told Mike Von Joel that anyone considering opening a gallery would have to be mad. In 1986, he was diagnosed with AIDS. The gallery closed soon after, and Robert died at his mother's house a year later. Fraser is often seen as the first star gallerist, and he was certainly one of the movers and shakers of the briefly swinging city.

The most improbable 1960s contemporary art dealer was Alex Gregory-Hood at the Rowan Gallery. A decorated colonel in the Grenadier Guards from a Warwickshire land-owning family, Gregory-Hood had faced the unusual choice of accepting a promotion to general or leaving the army to open an art gallery. He was a glamorous figure who drove an Aston Martin and wore kaftans when in the country. The gallery's first home was in Lowndes Street, a lonely outpost of contemporary art in Belgravia. Opening in 1962, the same year as Robert Fraser's gallery, the Rowan Gallery concentrated on British abstraction and included artists such as Phillip King, William Tucker, Sean Scully, Barry Flanagan, Paul Huxley and, later, Bridget Riley. Michael Craig-Martin recalls that Gregory-Hood took him on a 50/50 basis and 'sold nothing but remained doggedly loyal to me as he did to all his artists whom he defended like his country. There was a naivety about the Rowan Gallery that made it very attractive.'[19]

In 1968, the Rowan moved to Bruton Place – closer to the centre of the contemporary art trade – where Timothy Rendle designed the beautiful gallery space. That year, Gregory-Hood staged the first Warhol exhibition in London, a considerable feather in his cap. When the Bruton Place lease ran out in 1982, he went into partnership with Annely Juda and her son, David, to form Juda Rowan, which lasted until 1987. He needed her space, but they had a very different approach to life. While he enjoyed long, bibulous lunches with artists which started at 11 a.m., these were an anathema to her. James Mayor (who also had a brief partnership with Gregory-Hood) believes that Rowan, like many galleries, was sunk by rising property prices. The consensus was that Gregory-Hood was 'too nice a person to be a dealer'.

Annely Juda is a fascinating and courageous dealer whose commitment to abstraction was impressive. German-Jewish, Juda was born in Kassel in 1914, fleeing Germany in 1933 after her father was arrested. She spent the war in the Women's Auxiliary Service delivering food to people whose homes had been bombed out. After a spell back in Germany, she returned to London in 1955 and gained experience of art dealing by working for Eric Estorick at the Grosvenor Gallery. She opened the Molton Gallery in 1960 and became known for exhibiting Russian constructivism, the bauhaus and de stijl, and showing artists such as Kandinksy, Malevich, Mondrian, Tatlin, Gabo, Lissitzky and Rodchenko, displaying a taste which was rare at the time in England. The first exhibition was *Now Open: Important Paintings of the 20th Century and Young Artists*.

The most striking aspect of her exhibitions was showing Russian avant-garde work at a time when people still were unsure about authenticity or, as the great collector George Costakis put it, 'Who were the generals and who were the captains in the story'. Juda opened the Hamilton Gallery in 1963. In 1968, she changed the name and reopened as Annely Juda Fine Art in a warehouse space on London's Tottenham Mews before ending up in Dering Street with her son, David. Her rigorous abstract art exhibitions in the 1970s were much admired, particularly a series called *Non-Objective World*, showing Picabia, Jean Arp, Sandor Bortnyik, Malevich, Mondrian, Kandinsky, Helion and Pasmore.

During the 1960s, two short-lived galleries lit up the London sky. The Indica Gallery (1966) in Mason's Yard was a brief but brilliant firework under John Dunbar, then married to Marianne Faithfull. Very Swinging Sixties, it was patronised by Paul McCartney and by all accounts was heavy with drugs. The gallery's most famous artist was Yoko Ono. It prefigured the galleries of today by combining a book shop, art gallery and café. Another intriguing arrival was Signals Gallery (1964–6) on Wigmore Street, run by Paul Keller and described as 'The fulcrum for Avant-garde, international art practices from Europe and Latin America that centred on multi-disciplinary, participation, kinetic experiment and political engagement.'[20] Signals became a centre for experimental international artists. Neither gallery prospered, but both deserve recognition.

Helen Lessore took over the Beaux Arts Gallery in 1950s. An artist and former pupil of Henry Tonks, the great teacher at the Slade, she was at first unable to afford any famous names. Looking for new talent, Lessore nurtured 'kitchen sink artists' such as John Bratby, Jack Smith and Derek Greaves. The setting was unlike any gallery of its time, with simple floorboards and 'an informal and bedsitterish atmosphere'.[21] The Bruton Place premises had in fact been her husband Frederic Lessore's old sculpture studio and consequently had wonderful daylight. Frank Auerbach had his first proper exhibition at the Beaux Arts Gallery in 1956, and the show was described by David Sylvester as 'the most exciting and impressive one man show by an English painter since Francis Bacon in 1949'.[22] Lessore confessed that being an artist 'put me on the side of the artist, rather than of anyone else'. She supported Kossoff, Andrews and Uglow early in their careers. Richard Morphet, the Tate keeper, thought that 'her most important quality was moral; she was utterly incorruptible'.[23]

The Crane Kalman Gallery prospered in Knightsbridge. Andras Kalman, who arrived from Hungary via Manchester, was described as 'a character from a Dickens novel'. His distinctive theatre and film clientele included actors like Alan Bates, and Frederick Forsyth was another customer. Kalman was known for stocking two very desirable artists, Ben Nicholson and L. S. Lowry. They both had main dealers but liked Kalman and made an exception for him to handle their work. For Lowry, the Manchester connection was attractive. Kalman's influence lived on through Ivor Braka, for whom he was a mentor.

Ben Nicholson is an interesting example of the fluid relationship between artists and dealers. He started with Lefevre in 1935 and had an exhibition there virtually every year until 1954, when he got his big international break at the Venice Biennale. This precipitated a European tour of his work that finished at the Tate Gallery in 1955. At that point, Nicholson moved to Gimpel Fils because he liked the family and they promoted his international reputation, particularly in Switzerland, which was rich and a location where works could be sold in a tax-advantageous way. In 1963, Nicholson had his first show at Marlborough, where he admired the design of their publications, the quality of the photography and the way in which art was catalogued – few artists were as fussy about such things.

In 1969, Nicholson was on the move again, holding an exhibition of graphic work, mostly etchings, at Leslie Waddington. He had two more exhibitions at Marlborough in 1970 and 1971 but went back to Waddington in 1976 with a major show and never returned to Marlborough. Waddington put on several shows for Nicholson but was never given an exclusive contract, as the artist still enjoyed working with Andras Kalman. Waddington was given the new work, and Kalman was allowed to have small retrospective shows. When Nicholson died, he left instructions that Leslie should represent him. The things that mattered to him were his international profile, the quality of the catalogues, photography, reproductions, words and, finally, friendship. He clearly liked both Kalman and Waddington very much.[24]

One young dealer who didn't just sell American artists but had lived amongst them and went to their parties was Freddie Mayor's son, James. Blessed with good contacts and high vitality, he did not come of age until the 1970s but is an important link in bringing American art to public collections in Britain. Like so many others, he confessed, 'I owe everything to Bryan Robertson.' James was fortunate in being sent to work for Sotheby's New York, where he was one of the organisers of the mother of all contemporary art sales: Robert and Ethel Scull's celebrated auction in 1973. Returning to London the same year, he took over his father's gallery in Cork Street. His New York sojourn had served him well and he had the immeasurable advantage of gaining the trust of the doyen of New York contemporary dealers, Leo Castelli.

Mayor was to become Castelli's London satellite and opened with a Warhol exhibition of *Mao* screen prints. He was to put on the first commercial solo exhibitions in London of Robert Rauschenberg, Franz Kline and (notably unsuccessfully) Cy Twombly. He couldn't sell any of Twombly's works until he took them to Basel in 1973. When Castelli wanted to do a Lichtenstein bronze show, he did it jointly with Mayor in 1977. Mayor's relationship with Lichtenstein was to be particularly fruitful, and the gallery staged six London shows of the artist's work. James was staying with Roy and Dorothy Lichtenstein when the news came through that Picasso had died. He recalled someone from the press calling the artist for his comment and Lichtenstein responding: 'Well I guess we all move up a peg then!'

Mayor was lucky to have two great collections on tap, the Scull collection in America and the Penrose collection in Britain. One of his greatest coups was to sell Jasper Johns's *False Start* (1959) from the Scull collection to François de Menil. His triumph on this side of the Atlantic was handling the sale of Penrose's most famous Picasso, *Weeping Woman*, for £3.1 million to the Tate. Mayor particularly relished selling contemporary art to British provincial galleries, which could be a slow and painful but rewarding pastime. He sold Francis Bacon's *Portrait of Henrietta Moraes* (1965) to Tim Clifford at Manchester Art Gallery, and works by Lichtenstein, Wyndham Lewis, Picasso and Stanley Spencer to the Scottish National Gallery of Modern Art. Mayor came to know Andy Warhol and commissioned some animal portraits of dogs and horses from him, subject matter always appealing in Britain. When Mayor wrote the artist's obituary for British *Vogue*, he described Warhol as a ringmaster for performances around him, persuading otherwise conventional people to do unconventional things. In his view, Warhol was actually a rather clean-living voyeur who enjoyed other people's decadence. The Mayor Gallery in Cork Street continues to this day, still under the direction of James Mayor, a 1960s survivor if ever there was one.

Chapter 9

Furniture: Bond Street

'It was the high point of grand traditional taste before the auction houses took over.'[1]

Jonathan Harris

As with the picture dealers, the furniture trade at the beginning of this period is governed by Bond Street grandees, in this case Mallett's and Partridge, both now one with Nineveh and Tyre. In their heyday, these firms catered to the tastes of the tycoons of Mayfair and Belgravia, and the Americans who visited London each spring. As clients became more discerning about fakes, the market for traditional furniture was dominated by Christie's successful 'provenance' sales. The furniture trade witnessed seismic changes in taste and in the ethics of restoration. From the 1960s onwards, it also saw changes in geography, as its centre of gravity slowly shifted from Bond Street westward to Pimlico Road, as we will see in the next chapter.

In the mid-century, the writings of the preeminent dealer R. W. Symonds were paramount among furniture collectors. As the main reference, he provided the benchmark of quality in the field, emphasising Queen Anne and Georgian taste where materials, craftsmanship and condition mattered.* During the 1960s,

* The taste was typified by the collection of Colonel Norman Colville and Percival Griffiths.

the study of English furniture came into its own. The Furniture History Society, founded in 1964, published the *Dictionary of English Furniture Makers, 1660–1840* in 1986. Standards changed as academic research improved, and the auction houses began to catalogue items in greater depth. As with most art, when prices rose fakes appeared and collectors began to ask more questions and became more discriminating. In line with this, vetting committees for fairs also became more important.*

During the 1960s, the taste for Regency and neoclassical Thomas Hope furniture designs spread beyond the wealthy Brighton bachelors who liked a bit of theatre in their interiors. You can still glimpse this taste in old films. Victorian furniture and decorative arts were re-examined in the hugely important 1952 exhibition of Victorian and Edwardian decorative art at the V&A. Above all, there was a growing emphasis on provenance, context, authorship, precise dating and design sources.[2] By the late 1970s, interest in provenance gave ascendancy to country house material and we can follow this through the major sales at Sotheby's and Christie's. Serious buyers increasingly wanted a famous house or collector attached to their furniture.

'By appointment to Queen Mary' seems retrospectively to sum up Mallett's old-fashioned grand Georgian good taste, but during the 1960s the firm could still see themselves as tastemakers. Mallett's was the Bond Street destination for all lovers of English furniture.† Founded in Bath in 1865, the firm

* Later, Lucy Wood's comprehensive volumes on commodes and upholstered furniture in the Lady Lever Art Gallery (published from 1994 onwards) were the first to study furniture construction in such depth.

† The V&A possesses more than fifty pieces of furniture which once formed part of their stock.

moved to London in 1937. It was not until 1955, when Francis Egerton became chairman, that Mallett's began to assume the character that most people remember.

Egerton was the guiding spirit of Mallett's glory years. From the Welsh Guards (awarded an MC), he was destined for the Foreign Office but somehow ended up selling furniture. Described as impish and incredibly fastidious, he would look at a piece of furniture and purr that it was 'a dweam of beauty'. To friends, he looked and sounded like Horace Walpole and had a similar adulation for strong women, including Mary Anna Marten and Lady Margaret Fortescue, for whom he furnished grand houses. By contrast, Egerton rarely knew the names of his secretaries, and it was said that he failed to recognise Mick Jagger when he appeared in the shop and left to find favour at Partridge.

Egerton conceived the concept of supplying the whole house – everything down to lamps and porcelain. The firm even had Dudley Poplak, a pet decorator, to provide a 'Mallett look' characterised by brown furniture and panelled good taste, enlivened with decorative porcelain. One of the most enjoyable operations for the directors was what they called 'doing a van job', which entailed taking a tranche of furniture to a customer's house in the country or flat in Eaton Square to see how it looked. Although Mallett's panelled boardroom taste appealed to City financiers, they depended on the American clients who came over in May and June, and they sold 50 per cent of their annual turnover during those months.

What were their sources? Egerton would often buy pieces from Hotspur, the highly respected firm in Lowndes Street, and every week he would venture westward. His first call would be to the equally revered Rubin brothers at the Pelham Gallery in the Fulham Road. Pelham mostly served the trade, a filter between the country and the West End dealers. It was said that nobody in the trade would pass through London without visiting their shop. Francis Egerton's next port of call might be two galleries on the King's Road, 'Jeremy' run by Geoffrey Hill, an ex-RAF gentleman-dealer who knew exactly what Francis wanted, and David Drey, who came up with fun and quirky items. David Salmon remembers Egerton smoking a cheroot outside his shop on the King's Road at 9.30 a.m. every Tuesday morning and always found him charming to deal with.

On Wednesdays, Egerton would visit Kensington Church Street to see Jonathan Harris and the glass specialist Eila Graham. Once a year he drove his Bentley on a buying trip to the Continent, especially Italy, returning with a crate full of decorative items. Alongside this, there were the 'runners' who would bring things to the shop. These included Cecil Lewis and his nephew Danny Katz, 'Mac' McSweeney and the dealer 'Dick' Turpin, a big man with a drooping black moustache and a curiously high-pitched voice.[3] As David Salmon put it, 'Francis had the golden eye and taste that John Partridge didn't have; you would have never caught John down the King's Road.'[4]

In 1961, Mallett's established a second shop at Bourdon House, a beautiful eighteenth-century building on Davies Street, where they supplemented Bond Street with French and Continental furniture. This more adventurous stock dated from all periods. As Martin Drury told the author, 'Bourdon House was acquired by Mallett's to compete with Partridge. They were Oxford to our Cambridge, and we were desperate to even up with them.'[5] The shop was run by the rather peculiar combination of Sir Guy Holland, an English baronet flying under amateur colours, and the Polish connoisseur Andrew Ciechanowiecki, Mallett's expert in bronzes and French furniture. Ciechanowiecki would disappear to Paris to go shopping amongst the Polish dealer community there, but in the end he was never going to be a comfortable match at Mallett's and left to set up Heim. At this point, the fun-loving David Nickerson took over Bourdon House. One day, he was showing Fred Astaire and Bob Hope

1. (*above*) The night everything changed: Peter Wilson taking the Goldschmidt sale at Sotheby's on 15 October 1958.

2. (*below*) The directors of Agnew's in 1963 by Lord Snowdon. Sir Geoffrey Agnew is second from the right.

3. James Byam Shaw: the ideal of the scholar dealer.

4. Erica Brausen, who fattened up Francis Bacon for Marlborough to gobble up later.

5. Cuckoos in the nest: Harry Lloyd and Frank Fischer of Marlborough.

6b. Fraser handcuffed to Mick Jagger in Richard Hamilton's *Swingeing London '67*.

6a. Robert Fraser: the hippest dealer in London.

7a and 7b. 'The most beautiful art space in London': Kasmin and his gallery.

8a and 8b Julius Weitzner (*left*), who found more 'sleepers' than anyone else, paid £2,450 for Rubens's *Daniel in the Lion's Den* at Bonham's in 1963 (*below*).

9a and 9b Norton Simon (*left*) disrupts the sale of Rembrandt's *Titus* at Christie's in 1965 (*below*).

10. (*left*) Velasquez's *Juan de Pareja* made 2,200,000 guineas at Christie's in 1970, a record not beaten for ten years.

11. (*right*) Titian's *Death of Actaeon*, bought by Weitzner for 1.6 million guineas at Christie's in 1971.

12. (*left*) Christie's Impressionist triumph: Monet's *La Terrasse* made 560,000 guineas in 1967.

13a. Regency silver takes off with the Harewood service at Christie's in 1965 (*above left*).

13b. The legendary silver dealer Ben How (*above*) with her mastiffs.

14a and 14b
'A dweam of beauty':
Francis Egerton guides the Queen Mother (*right*) around Mallett's showroom (*above*).

15a and 15b. Pimlico Road bohemians: John Hobbs (*above*) and Christopher Gibbs (*below*).

16a and 16b. Mandarins: Julian Thompson (*above left*) and Giuseppe Eskenazi (*above right*) between them owned the Chinese market.

17a and 17b Disrupters: Tom Keating (*left*) and Geraldine Norman (*above*) made the art world look foolish.

a very large mirror, and the three broke into a soft shoe shuffle at the very moment Queen Elizabeth II appeared with Francis Egerton.[6]

Bourdon House had plenty of room, and the restoration workshops were rehoused there. Mallett's sold things in tip-top, polished condition, employing as many as fourteen craftsmen in various workshops around London. One of the most important was the gilder, George Thomas, to whom Francis Egerton would hand over a new acquisition with the instruction 'this needs surgery', which meant alteration. As a later director, Lanto Synge, reflected, 'in time the practice became wholly unacceptable'.[7] Although tastes were becoming simpler, Egerton's instruction was often to make items conform to his own more elaborate taste. The workshops were populated by cockney characters who were very direct. When the Prince of Wales visited, he asked one of them what it was like working at Mallett's: 'It's alright sometimes,' came the chirpy answer.[8]

As Lanto Synge recalls, 'when I came to Mallett in 1969 "old-fashioned taste" was rather frowned upon. It was represented by heavily carved, rather dark (almost black) mahogany furniture, sometimes upholstered with old, somewhat faded, and definitely dusty needlework, probably the original covering. These aspects were despised in favour of simpler lines, as in shapely, less carved cabriole legs, plainer veneered furniture, pale satinwood and elegant painted furniture.'[9] In fact, by the late 1960s, the firm was to find itself representing old-fashioned taste as the new delight in patina and eclectic items from around the world became part of the *mise en scène* of the younger generation of hip buyers. Every year at Grosvenor House, Mallett's exhibited a similar red lacquer bookcase decorated with gold chinoiserie from *c.*1720, which had once seemed exciting (they even sold one to the British Rail Pension Fund). As one furniture specialist rather unkindly put it, 'Mallett found itself stuck in dowager taste.'

The other Bond Street furniture titan was Partridge, which had a more Continental atmosphere than Mallett's. Cultivating the rich international residents in Grosvenor Square, they became the best-known purveyors of French furniture in London. Just after the war, Partridge moved into 144–146 New Bond Street, opposite Sotheby's, and turned it into one of the handsomest purpose-built galleries in London.[10] At the same time, Claude Partridge introduced the French element into their stock, a move which turned out to be inspired. The main figure of the period, however, is Claude's son John Partridge, a great enthusiast who held sway from the glory days in the 1960s until the firm's demise in the new millennium.

If a cartoonist was looking for a caricature of a grand art dealer it might have been John Partridge: charming, rounded and slightly overdressed with his jewelled tiepins, etc. One day I asked John Partridge who the greatest expert on French furniture was in London, 'You are looking at him' came the response. Not many would have agreed with this self-assessment, and many thought that his expertise left something to be desired, but he always had excellent deputies, like Wynn Bunford and Clifford Henderson.

John Partridge's big break came when he was twenty-seven. He acquired a third of the Alfred Chester Beatty furniture collection, the other two-thirds of which were bought by Paul Getty, Paul-Louis Weiller and Stavros Niarchos. In what he describes as 'A major turning point in my life', John saw his chance in a conversation with Peter Wilson about the Goldschmidt sale.[11] As a solution to Wilson's conundrum about what to do when the seven-lot sale was over, he asked, 'Why don't we do a joint effort and have a reception afterwards for everybody at Partridge's?' That evening, John sold enough of the Chester Beatty furniture to pay for 50 per cent of what he had spent on the whole collection.

Throughout the 1960s, Partridge sold equal amounts

of French and English goods. The Greeks in Grosvenor Square and Eaton Square, like the Livanos, Goulandris and Mavroleon families, bought his French commodes to place under their impressionist paintings. The Americans tended to prefer English, but when the New York collector Henry Kravis swapped his English wife for a French one, he too began to buy French furniture. At a time when French taste was sweeping down Fifth Avenue, Jayne Wrightsman and Henry Ford were major buyers of important French pieces. Partridge's biggest buyer by far, however, was Niarchos, who was furnishing the Hotel Chanaleilles in Paris.

In 1964, John Partridge innocently found himself at the end of a messy 'ring' story. On 8 November that year, the *Sunday Times* reported on 'The Curious Case of the Chippendale Commode', telling the story of a lot in a Leamington Spa auction sold to the ring for £750 and bought by Major Michael Brett, a council member of BADA. In the ring's knockout at the Clarendon Hotel, it sold for £4,350. The dealer who bought it took it to Partridge the next day and sold it on for a much higher price. Occurring at the same time as the Northwick Park ring scandal, the affair blew the lid off the practice.

Below Mallett's and Partridge, the top tier of furniture dealers included Blairman, Hotspur and Phillips of Hitchen, who all sold to each other and enjoyed close relationships. Hotspur were in Lowndes Street next to Belgrave Square. Robin Kern, the young dauphin of the firm, joined in 1956 after National Service with the Royal Army Pay Corps, and his brother Brian joined shortly afterwards. As part of his training, Robin made a Grand Tour of American and European collections with Jerome Phillips, the son of a colleague, demonstrating the genial nature of the furniture trade. Hotspur took English furniture seriously as works of art in the same way that the dealer Frank Berendt did with French furniture. They both had a small stock of very discriminating pieces which they sold to new money clients like

Lord Harris Lanto Synge thought that 'Hotspur were the best dealers in terms of authenticity'.[12]

Away from Bond Street and Pimlico Road, the main London streets for antiques were the King's Road, Fulham Road and Kensington Church Street. Many dealers, especially in porcelain, had gravitated to Kensington Church Street from Portobello Market. The most notable furniture dealer there was the scholarly Jonathan Harris. He had been to Trinity College, Cambridge, which at the time was unusual, as very few members of the furniture trade had been to university. After a short period at Sotheby's and then as a runner, he set up shop in Kensington selling a fastidious mixture of furniture and works of art. The measure of respect for Harris is that for ten years he was the furniture expert on the Acceptance-in-Lieu panel.

In a period of changing tastes, the fondness for oak furniture was kept alive by Sam Wolsey in Buckingham Gate. He supplied many museums and furnished houses like Athelhampton in Dorset for Robert Cooke, the MP and historian of parliament. Oak made something of a comeback in the 1970s, in no small part thanks to the arts and crafts revival and Cotswold money. Not all the great furniture dealers were based in London, and Jerome Phillips of Phillips of Hitchin (founded in 1884 as pawnbrokers) supplied the London trade. Perhaps the most respected provincial dealers, they supplied one of the most colourful of all furniture collectors, Judge Irwin Untermyer of New York, whom the Met director, Tom Hoving, described as 'a character out of Lewis Carroll'.[13]

It was the Christie's Furniture Department that in many ways was to be the market maker, with Sotheby's rivalling only in the field of French pieces. After old master paintings, furnitureas Christie's second most profitable department. They raised the game through the quality of their cataloguing and their scholarly expertise. In the 1960s, the department was run by Anthony Coleridge, who wrote the then standard account of Thomas Chippendale. Coleridge had one disadvantage as an auctioneer: he was short-sighted. During the Cecil Beaton sale in 1980, he was selling lot twenty, a small table of low value. His main bidder was a lady seated in the second row. Against her, as Coleridge relates, was 'someone standing in the gloom at the very back of the tent who appeared to be raising her arm aloft ... when the bidding reached 5,200 guineas and I was beginning to think it a bit strange, one of my colleagues came up behind me and whispered, "you are taking bids from a carved wooden statue at the back of the tent". It was a German baroque figure of a partially draped female nude.'[14]

The sale of treasures from Harewood at Christie's in 1965 was probably the first auction catalogue in which furniture was illustrated in colour, although the cataloguing itself was still perfunctory. Foremost among the items listed was the Chippendale library writing table, acquired by Blairman at a world record price of £43,050 for the regional museum Temple Newsam. In 1971, Christie's sold the dazzling furniture collection of the Detroit socialite and philanthropist Mrs Anna Dodge in London. Among the star lots was the Carlin table, bought by Habib Sabet, who had the Coca-Cola concession in Iran. Until the revolution, when he left the country, he had been building himself a replica of the Grand Trianon. The table was resold by Christie's in December 1983 and went to the Getty Museum. By this point, the auction catalogue entry had about ten times the amount of detail than it had in 1971, which reflects the new scholarly interest in furniture.

If Christie's furniture department raised standards and gained collectors' confidence, Grosvenor House Fair's vetting committee was equally important, especially after several fakes scandals rocked the trade and undermined collectors' faith. Opening its doors in 1934, Grosvenor House was an attempt by furniture dealers to bolster trade and encourage new buyers nervous of entering Bond Street shops. The fairs were great events, and there was a sense of occasion and fun. Royalty always made a special point of attending, especially Princess Margaret.

Taste, however, was changing towards the new decorator dealers of Pimlico Road. The high gloss and shininess of the gilt that was the hallmark of Mallett's and Partridge's reputation was going out of fashion. Increasingly, John Partridge found himself peddling the last remnants of a Rothschild taste for grand English eighteenth-century portraits and French furniture to clients like Ambrose Congreve at Warwick House. Partridge would not change, and the old business model of buying and restoring things no longer worked when everybody knew the price of everything. He wearily noted that the world was losing interest in the decorative arts: 'Young people do not want them; they are not interested in decoration. They like to have white walls. Young people hardly use dining rooms now … It is a different world.'[15] John's son Frank explained, 'we made money selling from country houses to the new rich; essentially we were restorers who added value to works of art through restoration and re-gilding, but by the 1990s the goods were beginning to run out'.[16] John Partridge took the company public in 1990, and for a while all was fine.*[17] However, family difficulties and the consequent takeover were a reminder of the struggles involved in running an art business to the tune of outside shareholders. It was no surprise when the company slipped beneath the waves shortly afterwards.

Mallett's had a similar fate. Despite retaining great experts like Lanto Synge, with Francis Egerton's retirement the firm

* Their profits for the difficult years 1993–4 were a healthy £1.76 million. John Partridge paid himself £93,250 that year, the same as Leslie Waddington but less than a third of what Richard Green was paid.

lost some of its magic. This coincided with the company going public in 1987. Mallett's did not attract easy shareholders. Mohamed Al-Fayed became a major investor who was not considered helpful, and he sold his shares to Lord Weinstock, an equally difficult personality. They cashed in the real estate and something went out of the business when it left Raymond Erith's specially designed galleries at 40 New Bond Street as well as Bourdon House. In the hands of investors, Mallett's became a 'problem company', seeming to lose direction just as tastes and clients were changing. Most people would agree that going public was a huge mistake for Partridge and Mallett's; it gave them different priorities and expectations that could not always be matched. The new shareholders wanted reliable dividends, and with the vagaries of the art market this was always going to be difficult.[18]

One dealer bucked the trend of closure. In 1975, Ronald Phillips moved the firm he founded in 1952 to Bruton Street just off Bond Street. In the beginning, Mallett's and Partridge were his best clients. Phillips's son Simon recalls, 'It was a typical ladder, of sales from a house sale in Yorkshire to a dealer in the Cotswolds to Fulham Road and thence to Mallett's.'[19] Ronald was happy to continue selling commercial furniture mainly to the trade, but when Partridge and Mallett's went under, Simon saw his chance. Young, keen to sell masterpieces and with strong nerves, he upped the number of staff and bought more expensive goods. His business model was to maintain a stock of very high-quality pieces such as the Ashburnham Place lacquer commodes. To some extent, Phillips benefited from the Richard Green syndrome of being the last man standing. Part of Simon's success is sheer hard work, not least of which is going out every evening and circulating to meet new buyers. He points out that the problem with Mallett's was that there was no family: for the staff it was a nine-to-five job, whereas for Simon Phillips it is all day, every day, including the weekends.

If Simon Phillips survives on the fringes of Bond Street, another great old firm has shown a remarkable adaptability. Blairman had a reputation for Regency pieces which (apart from Thomas Hope) began to fall out of fashion in the 1990s. Interestingly, the owner George Levy collected Victorian furniture at home and did not discourage his son Martin, the fourth generation in the firm, from moving towards arts and crafts. Martin Levy, who has a strong academic interest in furniture history, saw the writing on the wall for the big West End showrooms and repositioned himself, offering late nineteenth-century items from a top floor in Westminster which, as he put it, 'has enabled me to fight another day'.

The Christie's London team leadership passed from Anthony Coleridge to the scholarly Hugh Roberts, later director of the Royal Collection. John Partridge believed correctly that it was the auction houses which now attracted the buyers, especially for the great single-owner sales that Sotheby's and Christie's were still offering. The 1991 Messer sale at Christie's gained a lot of attention, representing the zenith of old-fashioned restrained English taste with a section on horology and Tompion clocks. Almost every piece was referenced in Symonds. The collection was less about historical pieces than about the quality of the individual work and the beautiful condition. The catalogue demonstrated a new level of sophistication: the endpapers were taken from Chippendale's designs, there was an erudite introduction and the first three lots were a selection of reference books. To the dealer Robin Kern, the sale was a turning point not because it represented old-fashioned taste, but because it was the moment when private clients and new buyers took over auctions and collectors like the American John Gerstenfeld secured the best lots.* Kern keenly felt the power of the new international buying world.

Christie's was to follow up the success of Messer with a series of dazzling 'provenance' sales at the end of the century. These included important furniture belonging to Lords Bute, Cholmondeley and Bath, and represented the thinning out of British country houses. The Cholmondeley sale of pieces from Houghton Hall had a French decorative arts emphasis and benefited from the nexus of Arab and Russian arrivals in London. These buyers competed against French couturiers like Hubert Givenchy and his millionaire compatriots: M. Pinault and M. Arnault. The single most important piece of furniture sold during the period was the Badminton cabinet auctioned at Christie's in 1990 and bought for £8.58 million by the collector Basia Johnson. Returning to the auction house again in 2004, it sold for £19 million to the Liechtenstein Collection in Vienna. Furniture was now achieving impressionist prices at auction.

By the millennium, many dealers, including Mallett's, Partridge and Apter Frederick, gave up and started selling their stock at auction. There are many theories about what killed the Bond Street trade. For some, the internet killed off showrooms and gave transparency to prices. Others blamed 9/11 and the end of regular American visits to London. Undoubtedly, John Partridge was right that there was a change in taste away from what was now disparagingly referred to as 'Brown Furniture'. Grosvenor House fair was a casualty. By the 1990s, Charlie Mortimer thought that the event had become too grand, 'like some huge top-heavy old battleship about to sink under the weight of its own self-importance'.[20] It finally closed its doors in 2009. Since the late 1960s, Pimlico Road was making the running.

* John Gerstenfeld first came to notice at the H. J. Joel sale of Childwickbury by Christie's in 1978.

Chapter 10

Pimlico Road

'Pimlico Road dealers were disrupters and they made furniture sexy. They were rock and roll and John Hobbs was a rock star.'[1]

Peter Holmes

Pimlico Road became a moniker for furniture shops in counterpoint to Bond Street. They were imbued with the spirit of the 1960s, the feeling of liberation from formality and irreverence for the bloodless good taste their parents and grandparents had admired. Unlike other markets that diminish or move to other locations, the journey of the furniture trade is one of colour and reinvention. A move towards colourful flamboyance evolved into something more atmospheric and eclectic – the essence of Pimlico Road taste. 'I remember the excitement,' wrote Martin Drury, 'of discovering the first shops at the far end of the King's Road over the bridge which were stripping paint from the cheapest pine furniture, wax-polishing it in the back of the shop and selling it in the front and on the pavement. That was in 1965.'* 'Pimlico Road' was very much about making furniture fun and interesting for a new generation who

* Interview with the author. Drury added, 'The best of the pine shops was called Avis Mostyn. She herself was an elegant, inscrutable woman with ash-grey hair and a consumptive complexion. The pine-furniture movement must have been a by-blow of Habitat.'

preferred a battered club look to the Partridge high gloss. It was often described as 'shabby chic', and it was where the market was heading in its appeal to a more hedonistic generation of British pop stars and couturiers.* The most conspicuous dealers were the exquisitely courteous Old Etonian *arbiter elegantiarum* Christopher Gibbs, and the talented, foul-mouthed John Hobbs.

Christopher Gibbs (1938–2018) was at the heart of London's 1960s *dolce vita* and managed to have a lot more fun than anybody else. Unlike Robert Fraser, he knew when to stop. Gibbs, always known as Chrissie, kept his life and friendships in good repair. With his armoury of charm, good looks, connections and fabulous taste, he was described in youth as 'part Montesquiou, part Beau Brummell and part Baudelaire'.[2] The writer James Delingpole noted at the end of his life that, 'with his silvery hair, well-cut but rumpled suit, and diffident, vaguely ecclesiastical air, Gibbs more closely resembles an Anglican dean than an acid-tripping ex-roué once known as the king of Chelsea'. Nobody epitomised the change in furniture taste from Bond Street to Pimlico Road better than Gibbs.

Although Gibbs was the most visible exponent of the new look, he was not the first or only practitioner. The decorator Geoffrey Bennison came before him, and to Brian Sewell, he epitomised:

> The cold formality of the Style Rothschild made warm and comfortable ... Geoffrey was a dab hand at such transformations; nothing looked new, rich things were used only if softened by age and use, value and importance were understated, rarity teased into knick-knackery and the knick-knack magicked into the curiosity; he conjured a sense of the nineteenth century with

* David Mlinaric told the author that the phrase was invented by Mirabel Cecil to describe his first studio in Tite Street *c.* 1963.

an inheritance from the 18th, lived-in and worn by generations of family, yet this marriage of Portobello Road and Bond Street (so to speak) was entirely new.[3]

The essence of the taste was to mix periods and styles with quirky objects in clever juxtaposition. If Bennison made Pimlico Road and created the look, Gibbs developed it with a romantic and scholarly twist and sold it to the younger generation of pop stars and their hip hangers-on among the super-rich.

David Mlinaric, who had a shop in Lower Sloane Street, recalled that:

> Gibbs was one of a handful of antique dealers, along with Piers von Westenholz and Geoffrey Bennison, whose shops opened in Pimlico and Chelsea in the 1960s; they helped to broaden tastes in interior decoration away from the impeccable Bond Street showrooms without immaculate stock and 'correct' arrangements of rooms, towards something tougher and more masculine, and, at the same time, less formal than conventional ... he embraced shabby chic with relish: his priority – the object, whether a piece of furniture or a picture, a book or a rug, matters more than its condition – has been inspirational in the creation and the arrangement of interiors.[4]

Gibbs's family had made its fortune in the guano industry (or, as the rhyme put it: 'Mr Gibbs / Made his dibs / Selling the turds / Of foreign birds'). Expelled from Eton for conning the local antiquarian bookseller into buying back its own stock, Gibbs set up his first antiques shop in Camden Passage in Islington with £10,000 from his mother. About the same time, he made his first buying trip to Morocco, coming home with 'rugs, lamps, djellabas, wall hangings and the name of the best hash dealer in Tangier'. Moving his operation to Elystan

Street, Chelsea, Gibbs ended up in Kasmin's old premises in New Bond Street. By now, he was a fixture of swinging London, reinventing the notion of dandyism by mixing clothes brought back from North Africa with British tailoring from Blades: 'You had to be monumentally narcissistic and have time on your hands, and just about enough money to do it,' he declared. His flat in Cheyne Walk provided the set for a marijuana-smoking party scene in Michelangelo Antonioni's film *Blow-Up* (1966).

The architectural historian John Harris describes a visit to Gibbs's shop:

> It was a memorable experience to leave the hustle and bustle of Bond Street, pass through that narrow darkened passage, to burst into the high, top-lit treasure house of salivation … I was always aware of how object answered object in many sensitive ways, and there was always what might be called creative rearrangements. I suppose the exhibit that evoked gasps from all and sundry was Lord Iveagh's sock cabinet from his bedroom at Elveden Hall, Suffolk; its drawers still containing an array of smelly socks wrapped around Sir William Chambers's designs for the cabinet, no less than the medal cabinet designed for Lord Charlemont at Charlemont House, Dublin.[5]

Style was the essence, and at Gibbs's you were as likely to find a Victorian bird cage as a vast Irish bookcase or a Grand Tour cabinet. If, on one hand, Gibbs loved objects designed by Pugin, he equally enjoyed displaying collections of humble walking sticks.

His fellow dealer, Jonathan Harris, commented that 'Chrissie loved original and amusing pieces, but he had an intellectual edge. He gained the trust of such collectors as Paul Getty Jr., who was his best friend, and unquestionably gingered up the market, taking it from the Swinging Sixties to the *World of Interiors* look. In this he was quite the opposite of John Partridge, who loved polish; Chrissie liked patina.'[6] Above all, Gibbs loved history and was a 'provenance fetishist', or, as he put it, 'I like intrinsically beautiful things, but if there's a yarn attached, that's a big plus.' 'I like things with a past and people, too,' he told one reporter, and it was suggested that, had he been on *Desert Island Discs*, he would have asked for a copy of *The Complete Peerage*.

Apart from Paul Getty Jr., whose cultural philanthropy he encouraged, Gibbs is most associated with the Rolling Stones. He furnished Mick Jagger's château in the Loire valley, placing Cusco School paintings against a rich multicoloured woven silk backing. Gibbs's carefree lifestyle is captured by the events of one night in the summer of 1968. Partying with Mick Jagger and Keith Richards at a South Kensington nightclub, at about 2 a.m. he suggested they adjourn to Stonehenge to watch the sunrise. Piling into Richards's chauffeur-driven Bentley with Marianne Faithfull, they went to eat kippers in Salisbury after watching the dawn. Keith Richards enjoyed staying in Gibbs's apartment on Cheyne Walk, partly for the endless supply of illicit substances and partly for his remarkable library of antiquarian books. Like Robert Fraser, Gibbs was present at Richards's house at West Wittering during the drug raid in 1967 that led to Jagger, Fraser and Richards receiving jail sentences.

Gibbs's romantic associations included a brief involvement with Rudolf Nureyev. On a professional level, he served on the committee that advised the V&A on the refurbishment of its British galleries, and Lord Rothschild put him in charge of furnishing Spencer House (with John Harris and David Mlinaric). Gibbs retired to his Tangier house in his beloved Morocco, and occasionally felt homesick for snowdrops.

*

Walking into the Hobbs brothers' shop in Pimlico Road was an extraordinary experience: you saw wonderful theatrical pieces which might be Russian, Scandinavian, Sicilian or Portuguese. They had ebullient and imaginative taste, introducing an exotic Continental element to Pimlico Road which made their gallery a cave of delight. Their father had a shop on the King's Road selling junk called 'Odds and Hobbs'. John was a youth of striking beauty who, Christopher Gibbs thought, 'strayed from a band of angels in a Quattrocento painting'. There the resemblance ended, because John Hobbs was a swashbuckling rogue. Although he was amusing, as one colleague put it, 'Unless living on a knife edge, he wasn't happy. John made the furniture seem glamorous and sexy.'[7] John's sidekick Charlie Mortimer summed him up well: 'In every generation of antique dealers there is one character who in the charisma stakes stands out head and shoulders above all the others. In this case it was John Hobbs. As John liked to say, very frequently, "You can't fucking buy charisma!"'[8] When he died in 2011, aged sixty-four, his main instruction was to 'make sure there are no fucking wankers at my funeral'.[9]

John and his brother Carlton, known as Carly, went in a few years from 'running' furniture to being major antique dealers. In between, John had a base in the King's Road Furniture Cave, where he played poker and backgammon, dressed in an ankle-length leather coat with the collar up. Mortimer saw John and Carlton as a strangely incongruous duo. Equally talented, Carly was shy and easy-going, while John was handsome and forceful. Martin Levy described Carly Hobbs as 'mysterious, brilliant, and charming; he had a great eye, flamboyant taste and sold masterpieces'.[10] In 1993, John and Carly split and John moved to a huge gallery in Dove Walk, where he was unrestrained by his brother. Mortimer gives an example of John's cocksure personality, describing how Elton John, a regular client, came into the shop one day with an Italian whom John failed to recognise as Gianni Versace. When the designer later sent a fax with a rather low offer on some pieces, John's answer simply read: 'Piss off you cheap Italian cunt'. This evidently amused Versace, and he even bought a few things.[11]

People, especially decorators, liked doing business with the Hobbs brothers because they were generous with their introductory commission, and they were always true to their word. 'Crucially,' said Mortimer, 'they secured the invaluable assistance of the Kent-based enigmatic genius (and I don't use the word lightly) restorer Dennis Buggins and his team of craftsmen. It cannot be overestimated just how critical Dennis's role was then. In the early days, antique Russian furniture was a firm favourite because it looked great, sold for fortunes, was very fashionable.'[12] Rudolph Nureyev was a customer, and John Hobbs loved asking Buggins to embellish pieces with ormolu mounts and exhibited a passion for medallions.

The end came when John refused to pay Buggins the £100,000 pounds he was owed. The restorer took his revenge by going public to the *Sunday Times* in 2008.[13] Buggins revealed not only that the furniture he handled passed through the auction houses as antique, but the extent of his work for John Hobbs. He had created or substantially re-created many 'antiques' using wood from old pieces bought from auction. Lord Hindlip at Christie's was heard to say, 'It explains why John Hobbs kept buying all those useless armoires for £2,000.' Another dealer recalled that 'John would occasionally pay a huge price for a piece at auction which would legitimise the rest of the stock.' It was suggested in the *Sunday Times* article that more than £30 million worth of antiques had been produced over a twenty-year period. Buggins, who employed at one time thirty workmen at a farm near Chilham in Kent, commented: 'Most people think 18th- or 19th-century craftsmanship is dead, but we've been doing it here.'

As prices rose, fakes began to bedevil the furniture trade during the 1990s. This was nothing new, but there was a different emphasis and scale. The problem during the 1960s and even 1970s had been the 'aggrandisement' of old furniture, as exemplified by Mallett's embellishments. The leading furniture restorer, Peter Holmes, was taken aback by the sophistication of the deceit of 1950s restorers. During the 1980s, there was a change of protocol amongst restorers, who became more scrupulous. The better ones began to turn away work which was unethical, and they were even placed on the Grosvenor House vetting committee. By the 1990s, as fakers became more ambitious, the mantra became 'Restoration bad; Conservation good'. Buyers, alarmed by fakes, now required higher scrutiny.

The most egregious case of forgery was two chairs allegedly from the celebrated set of St Giles chairs removed from the Earl of Shaftesbury's St Giles House, Wimborne St Giles. The chairs were sold at Sotheby's in 1996 for £837,500, doubling their pre-sale estimates. As Lanto Synge put it, 'the provenance fulfilled the expectation that this missing pair of chairs would turn up one day.'[14] The chairs were not, however, a missing part of the genuine set earlier sold by Christie's, but clever fakes knocked up to look like them. One dealer told the author, 'The difference between John Hobbs and most other amoral furniture dealers is that he faked it and sold it directly to the clients as genuine. Other dealers faked it and put the stuff into auction – of which the supreme example was the St Giles chairs.'[15]

The St Giles chairs episode, and later the Hobbs case, meant that a lot of American clients wanted their furniture reassessed. The scandals were bad for the market as people lost confidence, but it is remarkable how quickly they were forgotten, particularly in light of the Christie's 'provenance' sales discussed elsewhere. Fortunately, the fakes scandal coincided with the rise of scientific testing. The problem of fakes was not unique to London, and Paris was to be convulsed by the scandal of the Axa Boulle cabinets.[16] As the most influential non-dealer in the English furniture trade, Peter Holmes found himself an arbiter when Christie's Furniture Department brought him in to examine every important lot.

Chapter 11

Tribal Art: From Curiosities to Masterpieces

'We used to have great dealers like John Hewett and Lance Entwistle. The Sainsburys died, Hewett died. Sadly, all the great collectors have gone!'[1]

Hermione Waterfield

The story of the tribal art market in London is a farewell symphony. In 1945, Britain was well stocked with the spoils of empire brought home by soldiers, merchants, explorers and plant hunters, and they were assiduously collected by a group of pre-war enthusiasts like James Hooper. After the war, the world descended on London to feast on the dispersal of hundreds of years of accumulation via a series of major auctions in the 1960s and 1970s which effectively emptied the cupboard. With very few buyers in Britain, by the end of the century the market for tribal art relocated to New York, Paris and Brussels. While it lasted, the story was a colourful one, with some interesting characters.

The starting point for the British collecting of tribal art was Cook's three voyages to the Pacific Islands, Australia and New Zealand. The material returned from these trips, with its sublime simplicity, has always been the collector's Holy Grail. Despite this, following the tradition of General Pitt Rivers and

A. W. Franks, British collecting tended towards an ethnographic approach. This contrasted with the situation in France, where artists like Gauguin, Vlaminck and Picasso created a pictorial response to tribal artefacts in their own work. The influence of André Malraux's *Musée Imaginaire* and André Breton's collection of tribal art, small-scale antiquities and modern art created an influential taste. It was exported to America via Dominique de Menil (to be seen at her museum in Houston) and to Britain with Robert and Lisa Sainsbury, who endowed the Sainsbury Centre, Norwich. The combination of contemporary and tribal art helped to regenerate the latter market.

For anybody searching for tribal art in 1960s and 1970s London, Portobello Market was the place to go. Many significant dealers began their careers with a stall there, including James Keggie, Julian and Barbara Harding, Lance Entwistle and Peter Adler. For over thirty years, James Keggie's stall on Portobello drew in buyers like the influential Paris dealer Charles Ratton and the poet Robert Graves. Keggie was a generalist with a particular interest in the ethnographic evolution of weaponry and items of industry. Sotheby's was fortunate that Peter Wilson was interested in tribal objects and encouraged the formation of a separate department. It thus fell to them to stage the first single-owner tribal art sale in 1960, *A Highly Important Collection of African Art from the Belgian Congo*. Described as 'The Property of a Gentleman', the collection belonged to the Brussels dealer-collector René Whitofs.

Until 1969, when they held their first dedicated sale, Christie's put tribal art into any sale from furniture to antiquities. It was believed that London had poor expertise compared to Paris and Brussels, whence dealers would come to pick up bargains. The two outstanding Paris dealers, Pierre Verité and Charles Ratton, were as much a part of the London scene as any home-based dealer. Ratton was clever but could be tricky; he once refused to take back a mask from Robert Sainsbury, saying, 'Put it in a sale and I will buy it back cheaper.' Sainsbury never went back to Ratton after that. However, it was thanks to Ratton that Christie's was given an important collection to sell.

The sculptor Jacob Epstein broke with British ethnographic tradition, adopting a more French approach in his acquisitions. Today, his collection has classic status and Hermione Waterfield, Christie's first tribal art director, told the author, 'Epstein was the only one who did not mind starving for a good African statue.' Arriving at Christie's in 1960, Hermione found a young member of the furniture department cataloguing tribal art for the Epstein sale, which took place on 15 December 1961. 'Such a shambles,' as Waterfield put it, 'so full of mistakes that the corrigenda was longer than the catalogue!'[2] According to Waterfield, Ratton had expected that Christie's would make a mess, advising Lady Epstein to go there because he personally wanted many of the items and would be able to buy them cheaply. In fact, Ratton had already sold most of the best pieces from the collection to Carlo Monzino, and the Christie's sale was only the remnants.

If the Epstein sale didn't hit the top note, the Warwick Castle sale held by Sotheby's in 1967 certainly did. Madame Tussaud Ltd had just bought the castle and wanted to get rid of the tribal material. Much of it had been given to Charles Greville, the then owner of the castle, by the celebrated naturalist Sir Joseph Banks (1743–1820), and was therefore of paramount interest. The sale was something of a landmark and signalled a new awareness that these explorers' curiosities could be masterpieces. The star lot was the Hawaiian Male Figure which was purchased by George Ortiz, a compulsive, talented and often controversial collector who housed his remarkable collection of antiquities and tribal art near Geneva.

The giant of the period was John Hewett (1919–1994), who preferred to remain in the background. Something of an enigma, he was certainly a major player, attracting and maintaining the patronage of great collectors like the Sainsburys, Ortiz and many others. To John Baskett, he was an alarming figure who might have come out of a spy story. His parents were Plymouth Brethren and he briefly converted to Roman Catholicism. He joined the Scots Guards in 1939 and was posted to Italy, where he wrote poetry and started an art club. Opening his first shop in Richmond, he soon moved to Sydney Street off the King's Road, where he sold Egyptian antiquities alongside West African masks. Although Hewett often gave a rather detached and reclusive impression, this was deceptive.

Peter Wilson, who acted as best man at Hewett's wedding, made him a consultant at Sotheby's on both antiquities and tribal art. Cyril Humphris recalled Wilson's closeness to Hewett, whom he likened to D. H. Lawrence, on account of both his beard and his sensitivity. 'On visiting his little place,' Humphris recalled: 'There would not be very much on view, then he would open the cupboard and there would be an ostrich egg with a pre-dynastic figure. He was a man of mystery.'[3] There was a touch of the guru about Hewett, and Peter Adler recalled the change from groomed Guardsman to long-haired hippy. Hewett had discovered the band The Who, particularly, *Tommy*, and fell under the influence of youth culture. He knew how to handle very difficult clients and was also a favourite with the 'runners' because he took what they offered him and did not cherry pick.

Hewett met Robert and Lisa Sainsbury, his best clients, through Peter Wilson. He had an enormous influence on their collection and, as Waterfield describes, 'he introduced them to Polynesian art, Mediaeval, Cycladic and antiquities in general, and intimated to most other dealers that they would have to approach the Sainsburys through him'.[4] A good judge of character, Hewett knew exactly how and when to show an object. He knew that George Ortiz, for instance, liked to rummage around when his back was turned, so he would hide things for George to find and then demand to buy. One day, Hewett came down the stairs with the Benin bronze head that Ortiz was to call 'Bulgy Eyes'. It became one of the stars of his collection. Hewett was equally a master of saleroom psychology and would often get up and leave before the item he wanted to buy came up, remaining hidden from view but bidding from where the auctioneer could still see him. Edward Lucie-Smith said that 'he made objects available and he endowed them with magic'. Robin Symes, his protégée in antiquities, had the same gift.

Hewett's most questionable role was in the dispersal of the Pitt Rivers Museum at Farnham. 'The deal involving the Pitt Rivers Museum in Farnham is labyrinthine and mired in secrecy,' opined Nicholas Shakespeare.[5] The museum was housed in a converted farmhouse and school on the Pitt Rivers estate at Farnham in Dorset. Formed by the Victorian polymath and collector General Augustus Pitt Rivers, the collection included 250 works of art from the Benin expedition of 1897. By the 1960s, the family situation was singular. George Pitt Rivers, known as Captain George, the then owner, was a British fascist. Stella Lonsdale, his mistress, was thirty years his junior and had been suspected of being a British spy and incarcerated in Paris during the war by the Germans. She was described as 'a large, striking figure with a wicked sense of humour with a streak of white through her dyed black hair'.[6] Stella took George's name although they were not married, and after his death she married her much younger driver.

As George was very mean and Stella was a big spender, particularly as she had a house in France and a French lover in Marseille to maintain, she took matters into her own hands. There were rumours of lorries arriving late at night to empty the museum, and items being replaced with copies.* When Captain George died in 1966, it was impossible for his heirs to unravel what had happened. Stella inherited substantial property and evidently

* One of these copies appeared on the cover of a Sotheby's sale in 1976.

did a deal over the museum, but the details are unclear. Captain George had originally approached John Hewett to make an inventory of the collection, and he began to visit regularly. One thing leading to another, the museum was closed to the public and its contents were put into storage. Although Peter Wilson envisaged a one-owner sale, this never took place, and Stella preferred to sell individual pieces below the public radar, and that of the Export Reviewing Committee. John Hewett acquired much of the Pitt Rivers material privately and, as one dealer put it, 'There was a good deal of chicanery around the sales.'[7] Many artefacts were sold during the early 1970s through an offshore company based in Ireland, and Hewett admitted, 'I sold pieces to the US, to France, everywhere save in England.'[8] Some of the material was passed through Sotheby's as 'The Property of Stella Pitt Rivers', but most of it, particularly the Benin bronzes, was sold directly to collectors like Robert and Lisa Sainsbury, and one of the best pieces was Ortiz's 'Bulgy Eyes'.

When George's son Michael realised that the museum was being broken up and sold behind his back, he strongly objected. Stella also soon discovered that Hewett was selling the pieces on for vast multiples of what he paid her. As Peter Adler told the author, 'John Hewett became *persona non grata* after a piece bought for nothing from Stella made a fortune at Sotheby's. She was furious and John was no longer welcome, so he was forced to get Bruce Chatwin and myself to do his work for him.'[9] Chatwin was one of Hewett's protégés at Sotheby's.* An absurd charade followed in which Adler and Chatwin, both in their early twenties, were, as Adler recalls, 'Set up as wealthy collectors and dispatched by Hewett to Farnham in suit and tie to buy pieces. Since we were so inexperienced, we kept having to go to the village telephone box to ring John to ask how much

* Chatwin briefly headed what was then called the Department for Antiquities and Primitive Art, later renamed Tribal Art.

to offer on certain pieces. We both had a bit of a crisis of conscience about it. John was asking ten to fifteen times the price he was offering Stella.'[10] Chatwin liked to say that he fell out with Peter Wilson after a blazing row over Wilson's own private dealings over the sale of the Pitt Rivers collection, but he may also have had a retrospective sense of guilt about his own part in the debacle.

In 1978, George Ortiz, whom Chatwin dubbed 'Mighty Mouse' on account of his being diminutive and hyper-energetic, was obliged to have a sale of his own after his daughter was shockingly kidnapped (and later rescued) in Switzerland. The Ortiz sale at Sotheby's was the most important of its type brought into London. Collectors came from all over the world, and the sale attracted an audience well beyond the Brussels and Paris dealers. The items were presented by Peter Wilson as great works of art, many of them illustrated in colour in the catalogue. The fact that Wilson took the auction himself was an indication of its importance.* The prices were astonishing and represent the high-water mark of the London tribal market. Lot 232, a Hawaiian Figure, sold for an unprecedented quarter of a million pounds, and lot 119, the Pentecost Island Mask, made £180,000.

Christie's answer to the Ortiz sale was the auction of James Hooper's collection in five sales from 1976 to 1980. Hooper spent his life working for Thames Conservancy and supplemented his modest income by dealing to satisfy his enormous appetite for buying tribal art. His collection had sections devoted to Africa, North and South America, the Pacific Islands and, particularly, Polynesia. Hooper had opened his museum at Arundel in the 1950s, and Pathé even made a rather condescending film

* The Sotheby's 1972 winter catalogue had its first colour plate of a tribal piece: the Rarotonga Figure of a God, which had been with the London Missionary Society.

about it. The Christie's sale was superbly catalogued by James's young grandson, Steven Hooper. The five auctions totalled £2.8 million and put the Christie's Tribal Art Department on the map. The department now had William Fagg, the greatest expert on African art, as a consultant. When Fagg was about to retire as keeper of ethnography at the British Museum in 1975, Hermione Waterfield suggested to Peter Chance that Christie's make him a consultant. 'How much will that cost?' asked Chance, 'Only air-fares for visiting museums and collectors in America,' replied Waterfield.

Alongside Sotheby's and Christie's, Phillips started selling ethnographic material at their Salem Road saleroom in Bayswater. However, the cataloguing left much to be desired, and typical was lot 141 on 2 July 1976 (estimate £60), catalogued as 'South American Indian Ceremonial Cloak Collar'. Julian Harding, a knowledgeable dealer-collector, recognised it as a Tahitian piece from Cook's voyages and bravely took it up to £4,000, but it sailed on to reach £8,800 and ended up in Alistair McAlpine's collection. Harding estimated that about 80 per cent of all the important sale items left Britain.

Many shops sold tribal art in London, and it is worth mentioning some of them. William Ohly ran the Berkeley Gallery on Davies Street. Founded in 1941, the gallery sold a mixture of antiquities, Middle Eastern and tribal art (Pre-Columbian, Oceanic and African), alongside contemporary art. Although Sydney Burney is said to have been the first London dealer to exhibit modern art alongside tribal art in London, it may equally have been Ohly. He was born in Hull and lived in Germany between the wars. Horrified by the Nazis, he returned to London in 1934 and set up as an artist. Ohly's gallery staged several exhibitions of modern artists (Duncan Grant and Matthew Smith), but his tribal exhibitions were groundbreaking: they included, in 1947, *Art of Benin*, and in 1954, *Europeans Seen Through Native Eyes: Sculptures from Africa, Asia and America.*

John Hewett's main rival was Ralph Nash, a Jewish émigré born in Berlin, who was more sophisticated, had a better education and was even more secretive than Hewett. Nash extracted items from museums, including, it was even rumoured, the British Museum, by offering to swap pieces. One of the first to promote the 'masterpiece' concept, he sought to remove tribal artefacts from the 'curiosities' cupboard and would speak of them as 'The Art of Man'. Nash, who looked like Einstein, had a massive turkey neck, thyroid eyes and was a great flatterer. When this was pointed out, he said, 'yes but people still believe at least 10 per cent'. More international than Hewett, he was equally at home in Paris and Brussels. Nash was a big buyer at the Warwick Castle sale and acquired the celebrated Hawaiian Figure from the Ortiz sale.

Herbert Rieser's gallery near Marble Arch was described as more of a club than a gallery. Of South African German-Jewish parentage, Rieser was unusual in knowing Africa, and he would source his material there. One of his most engaging exhibitions was a selection of spoons from around the world. Rieser liked quirky things and was as delighted by a Polynesian fishhook as a Torres Strait mask. David Attenborough described him as 'sitting hunched in his chair, wreathed in cigarette smoke – as well as a drink brought out with a grin from behind his chair'.[11]

In the 1970s, Lance Entwistle started with a stall at Portobello selling African and Oceanic art, mostly material he had bought at Sotheby's. John Hewett encouraged him until he discovered that Entwistle was in contact with George Ortiz. Hewett was good at harbouring grudges, and after that Entwistle was cast into darkness. Entwistle dealt with his business partner Anthony Plowright from a flat in Hamilton Terrace and then in South Audley Street, but found that there were no clients in London. Instead, it was, as he put it, 'a market of true believers', mostly dealers. However, Entwistle, who was by now working with his American wife Bobbie, found a receptive audience in the USA and became a bridge between the European and American markets. His business model was based on knowing who wanted which piece and specialising in being a middleman for sales between museums and private collectors. Finding that all his clients were either in Paris or America, Lance Entwistle decided to leave London and base his business in Paris in 2004, banging the final nail in the coffin of the British market. At the same time, Sotheby's and Christie's ceased having any more major tribal sales in London.

Although Christie's South Kensington kept the story going until it closed in 2017, Hermione Waterfield described the London scene in the new millennium as:

> A desert! You need good dealers for collectors, and you need good collectors for dealers. The whole thing's just died off in London. We used to have great dealers like John Hewett and Lance Entwistle. The Sainsburys died, Hewett died. Sadly, all the great collectors have gone! London, no, England is a desert. It's curious, it's a chicken and egg situation. But there are no eggs to hatch. No hens to sit on them.[12]

Chapter 12

European Porcelain

'It is not just taste that has changed but the gap between the good and the average; everything has gone to the top of the pyramid.'[1]

Errol Manners

Anyone who ventures into the world of porcelain becomes aware of a vast network of dealers, markets, fairs, societies, exhibitions and scholarly activity far beyond the scope of this chapter, which can necessarily only deal with a few highlights. The porcelain dealer Errol Manners offers a crisp summary:

> The big London story in Western Porcelain was the rise of Hanns Weinberg of the Antique Porcelain Company, who bought so many of the most expensive lots and drove up the prices by abstaining from the cosy local 'ring'. It was a sad day for London when he suddenly removed most of his stock to New York. The emergence of Robert Williams from his mother's antique shop in Eastbourne to take on the big boys in London was the next chapter.[2]

Before the war, Chelsea 'gold anchor' wares, inspired by Sèvres rococo, were the most sought-after porcelain, as exemplified by Lord Bearsted at Upton House, who formed a large collection from the Chelsea, Bow, Derby and Worcester factories. In the same period, the influence of arts and crafts studio pottery created a fashion for simpler ceramics, which found favour among young collectors. A number of American collectors maintained the taste for French or German grand royal makers, and their annual London visits in the summer would be the vital selling time for these objects. For Hugo Morley-Fletcher of Christie's, the most discriminating post-war British collector in the grand manner was Harry Hyams: 'I can't think of anyone with better taste or more understanding of what he was buying.'* His collection (now belonging to a trust), housed at Ramsbury Manor in Wiltshire, has yet to be unveiled.

During the 1950s, there were said to be a hundred porcelain dealers in Greater London, about thirty of whom were specialists. Territorially, they tended to graduate from Portobello Market and congregate in Kensington Church Street or the Brompton Road, with a smattering in the West End around Bury Street. Jim Kiddell at Sotheby's was the leading auction house expert, although he was to be challenged by Anthony du Boulay at Christie's. Kiddell and du Boulay were friends who frequently had tea together to discuss objects and sales. Their genuine friendship and shared interest could not have been further removed from the later collusion between the senior executives over the buyer's premium. Kiddell was a much-respected figure who formed a 'Black Museum' of fakes, used to educate young trainees. It famously deceived a thief who stole most of the collection, and Kiddell had to start again. He would surprise and occasionally upset customers at the Sotheby's counter when he told them their piece was a fake, suggesting they take it to Phillips for confirmation before bringing it back to him, when he would pay £5, far above its true value.

Items at Sotheby's and Christie's often came from European collections and would be sold back to the Continent, where

* Hyams used an agent called Woolett, a dealer in Wigmore Street.

there were more collectors. This was partly because Europe never managed to develop an auction centre, and so London gained the market *faute de mieux.* Sotheby's and Christie's London sales were cosy affairs held around the 'horseshoe' table. As a measure of respect, Robert Williams was given a place among the reserved seats on the 'horseshoe' at Sotheby's, a jealously guarded privilege of the established dealers. He drew a map of the horseshoe in the 1960s with the caption: 'Woe betide any foreigner who dared to take the seat of any of the "Reserved" ones!! OUT. OUT!!'[3] The horseshoe – not confined to porcelain sales – made a glamourised movie appearance in *Octopussy* (1983), when James Bond bids for a Carl Fabergé Easter egg.

At Jim Kiddell's funeral in 1980, the collector T. D. Barclay turned to Anthony du Boulay, saying, 'Now you chaps have got this wrapped up', and he was right about European porcelain, although not Chinese, as we shall see. Described by a colleague as 'elf-like and excitable', du Boulay joined Christie's in 1949 and had to teach himself when faced with cataloguing a sale of Nymphenburg figures by Franz Anton Bustelli from Baroness van Zuylen van Nyevelt's collection, held the following year. As in so many other areas, Sotheby's tended to attract overseas and trade consignments and Christie's the country house goods. Typical of this was the Christie's landmark sale of Sèvres from Harewood House in 1964.

Du Boulay scored a considerable coup in 1958, when he persuaded his directors to recruit Hans Backer as their European porcelain consultant. Originally from Dresden, Backer was a dealer in Meissen porcelain who had established his business in London before the war and attracted the patronage of Queen Mary. Christie's top brass was initially resistant: 'We can't have a dealer,' cried Jo Floyd, 'nor a foreigner,' blurted Guy Hannen.[4] However, not only was Backer a first-class expert, he also recruited Geza von Habsburg as senior European representative for Christie's. A major figure in the firm's development, Geza had spent the war in Cascais in Portugal, where he knew all the exiled foreign royalty. One of them, King Umberto of Italy, gave Christie's his peerless collection to sell, which had a probable provenance of Augustus the Strong, the royal founder of the Meissen porcelain factory. Christie's sold the collection in the Richemond Hotel in Geneva in 1968, the first sale by any British auction house on mainland Europe.

In 1969, Robert Williams could claim that 'London is still the world's biggest [porcelain] market but there is increasing competition from the Swiss and the Germans.'[5] In fact, Christie's developed the Swiss market by locating sales there. Hugo Morley-Fletcher, Anthony du Boulay's younger colleague, correctly believed that he could find better material from the Continent and thereby broke the firm's country house dependence. Important sales of Meissen and Sèvres were sent to Geneva, having been catalogued in London. It was two sales with a Rothschild provenance that were to propel prices to a new high. First, the 1977 Sotheby's sale at Mentmore Towers in Buckinghamshire attracted new buyers for such desirable items as Marie Antoinette's ice pails. The following year, reversing their usual policy, Christie's brought the Edmond de Rothschild collection from Château de Pregny near Geneva to London for sale. Full of superb Nymphenburg and Meissen, it was the first porcelain sale to make more than £1 million.

The establishment of the French Porcelain Society in London in 1984 was a great boost for Sèvres collectors. It was founded by the dealer Kate Foster (later Lady Davson), their first president, alongside such scholars as Sir Geoffrey de Bellaigue, Surveyor of the Queen's Works of Art, and Rosalind Savill, director of the Wallace Collection from 1992. The treasurer of the society for many years was Adrian Sassoon, who has placed more Sèvres dinner services than most. Beginning as a receptionist at Sotheby's, Kate Foster had met the extraordinary collector Rudolf Just in Prague while on a study tour to Dresden. Just became immortalised by Bruce Chatwin, Foster's former colleague, in his book *Utz*. Just's collection was sold at Sotheby's Olympia in December 2011. They produced a two-volume catalogue in a boxed set which included Chatwin's novel. In 1974, Foster left Sotheby's to set up her own business and became one of London's preeminent dealers in European porcelain.

Carl Dauterman described Hanns Weinberg (1900–1976) as 'A Titan among the post war enthusiasts'.[6] Weinberg 'was so much more enterprising, so strong in his actions, so bold in his purchasing and so sage in his advice to collectors'.[7] Born in Dortmund, Hanns Weinberg trained to be a lawyer before he was detained by the Nazis. His family escaped Germany in 1938, arriving in London the following year, and were interned on the Isle of Man during the war. With no experience but unquenchable self-confidence, Weinberg set up the Antique Porcelain Company in Bond Street in 1946 and started to advertise in *Apollo* magazine. It was said that, after his experience of Nazi Germany, he only wanted to own movable goods and he always rented his property.

Weinberg immediately made his presence felt in the market and had, as one colleague put it, 'the courage of the devil'. As noted previously, he refused to join the ring and became known for paying very high prices at auction. Part of his talent was to attract the rich who bought French furniture and wanted some porcelain ornaments to garnish it. There were many dealers who were more scholarly, but none had more chutzpah. As the Lord Duveen of porcelain, he never liked to be outbid, and he asked prices that nobody else would have dared to utter. His first exhibition in 1951 attracted Queen Mary, and two years later the young Queen Elizabeth II visited his coronation exhibition of porcelain and renaissance jewels. The Duke and Duchess of Windsor bought porcelain pugs from him.

When the historian Sir John Plumb decided to form a collection, he was taken by Lord and Lady Cholmondeley – themselves collectors of Sèvres – to meet Weinberg. To Plumb, it was, 'The most exciting porcelain shop and the most expensive in the world … we were greeted by the owner, Hanns Weinberg, thin, pale as death with a wolfish, almost frightening smile: his eyes were heavy lidded, slightly bulbous, liquid brown, mocking, intelligent.'[8] Plumb found the shop 'amazing, case after case of dazzling porcelain, Meissen, Nymphenburg, Capodimonte, Chelsea, Worcester, Bow and what seemed like an entire floor of Sèvres'.[9] Whenever a price was mentioned, Plumb felt faint, but he was already hooked and became a client until Plumb found Hans Backer more scholarly and cheaper. Weinberg didn't like bidding on behalf of clients at auction because it restricted him; instead he offered them first refusal after a sale. When Plumb made other arrangements, Weinberg ensured that he was always outbid. Described as 'utterly ruthless', Weinberg appeared to have a low opinion of his fellow dealers and made sure that he always won.

Weinberg opened his New York shop in 1957 and left the London branch in the hands of his daughter, Rotraut. As Plumb put it, 'London had become too small for Weinberg, good for auctions, not so good for sales.'[10] Alongside New York he also opened a branch in Zurich, where Charlie Chaplin became a client. In America he attracted great collectors like Nelson Rockefeller, Jayne Wrightsman and the cream of Hollywood. Anthony du Boulay once visited Weinberg in his New York premises and remarked on the splendour of the pieces. This caused Weinberg to turn to his assistant, instructing her to 'double the prices!' His success was based on large, well-stocked shops in prestigious locations and his advice to collectors was simple: 'buy the best you can afford'.

Robert Williams, always known as Bob, was considered an easier figure to deal with than his older rival Weinberg.

Described as garrulous and generous, Williams didn't suffer fools, and could be quite tyrannical as a dealer, demanding that you buy something. He was born in Flint in 1923 to Winifred, a formidable mother, who set up a porcelain shop in Eastbourne. Once described as 'Queen Mary with a hint of a cockney accent', Winifred was self-taught and prospered selling Worcester, Derby, Meissen, Chelsea, Bow and Chinese *famille rose*. George Savage, the author of several authoritative books on porcelain, worked for her. After being demobbed from the army, Bob joined his mother. He trained under her eye and by visiting all the great European collections. During the 1970s, Williams moved to a small shop at 3 Bury Street in London's West End, very different from the magnificence of the Antique Porcelain Company. His premises were directly beneath Hugo Morley-Fletcher's Christie's office. When a new catalogue arrived, Hugo's assistant would call Bob's, who would go into the street and catch the catalogue from out of the window. In Bury Street, Williams put on groundbreaking exhibitions of Derby porcelain, *The Kakiemon Influence on European Porcelain* and a show on smaller French factories.

Williams believed that you could realistically only look after about ten collectors with differing collecting interests. Like Weinberg, he could be coy when asked to bid for a client: 'I'm going to buy that myself but will let you know if I'm successful.' He would only add 10 per cent on top and was considered very straight. Williams attracted young American collectors like the banker Robert Pirie, but his best clients were George and Helen Gardiner from Toronto. In 1977, Helen spotted a yellow ground Meissen cup and saucer in his shop window, which turned out to be part of a fifty-piece service. Having admired it, she uttered the dread words 'I will ask my husband', the kiss of death, as every antique dealer knows. Despite this, the next day George Gardiner appeared and their purchase was the beginning of a great passion. By 1984, they had opened their own museum of maiolica, baroque porcelain and scent bottles in Toronto. The most trying part of the relationship for Bob was staying with the Gardiners, since he was a gourmet. George had the franchise for Kentucky Fried Chicken, which he insisted on eating.

As we have seen, the training ground for porcelain dealers was Portobello Market. Among the many distinguished dealers who once had stalls there were Errol Manners, Brian Haughton, Jonathan Horne and Klaber & Klaber (Betty and Pamela Klaber). Jonathan Horne began in Portobello by selling pottery from the back of a car. From there he moved to a shop in Kensington Church Street, where he became the dominant force in English pottery from the late 1970s.

Errol Manners made a similar journey, spending six years at Christie's before he opened a stand in the arcade at Portobello in 1986. He handled items mostly priced between £65 and £400, since in those days £1,000 was about the limit of what could be sold at the market. The profit margins in porcelain were very low, because the collectors and dealers were fiendishly knowledgeable and the sheer number of them made it a highly competitive world. The average mark-up was about 30 per cent, with some dealers prepared to negotiate down to 10 per cent if necessary. The profit usually came from knowing just a bit more about a piece than somebody else, and being able to pin it down. Manners 'soon realised that putting on a mark-up was the wrong way to think, but rather to think in terms of what something was worth to one of your clients'.[11] In 1988, he opened his shop in Kensington Church Street, where he still specialises in seventeenth- and eighteenth-century European porcelain.

The Queens of Kensington Church Street were Betty and Pamela Klaber. Pamela's story is inspiring. She left school at the age of sixteen to attend the Study Centre course at the V&A under the formidable Erica O'Donnell. With her mother they started a business under the name of Klaber & Klaber in 1972, specialising in eighteenth-century English and European porcelain and enamels. The Klabers had a Saturday stand at Portobello and another in the Antique Hypermarket on Kensington High Street. They became members of BADA when Pamela was just nineteen. In 1974, they opened their tiny first shop in Hans Road next to Harrods, where they stayed for seven years. They would spend several days a week travelling, searching for stock and at the end of the week they would display their discoveries. Both collectors and dealers alike knew this, and they were always busy. Over the years they found clients by exhibiting at fairs around the country and advertising regularly in the *Antique Dealer* and *Collectors Guide* magazine.

In 1984, the Klabers opened a much bigger shop just off Kensington Church Street, where they remained for eighteen years. They rented the top floor to Geoffrey Godden (referred to by those who knew him as God), the great authority and author on English ceramics, who sold nineteenth-century porcelain. Proximity to so many other dealers helped the business, as collectors enjoyed walking up the street, which was like going to a fair. The Klabers always made sure they had whales and sprats for all pockets. Half of their stock was English and the other half European, with a speciality towards the early French factories. They staged a noteworthy series of exhibitions with scholarly catalogues including *Oriental Influences on European Porcelain* in 1978.

Pamela Klaber stressed the importance of fairs and exhibited at Burlington House, Chelsea, Bath, Cheltenham and Harrogate. They put their name down for the preeminent London fair, Grosvenor House, which had a long waiting list, and finally exhibited there in 1979, the year of the maids' strike at the hotel, which at the last minute cancelled the event. Performing a quick turnaround, they held an exhibition of the many pieces specially earmarked for the Grosvenor House

Fair in Hans Road. The Klabers both became chair of the fair's ceramics vetting committee. The decorative side was important to them, and at the grander fairs they would stock objects that furniture collectors wanted to put on their commodes as well as decorative botanical painted wares. Princess Margaret enjoyed buying porcelain and, at a preview evening of the Burlington House Fair, she arrived on their stand. Spotting an eighteenth-century English enamel box with the caption 'May the crown of Britain never tarnish with sedition', the Princess confided 'I would like to give that to my sister', adding 'But please don't tell her that I'm buying it', which greatly amused them, as they felt they were unlikely to run into the monarch.

According to Pamela Klaber, the game-changer was the International Ceramics Fair and Seminar, founded in 1982 by the dealers Brian and Anna Haughton, together with Len and Yvonne Adams. Initially held at the Dorchester Hotel before transferring to the Park Lane Hotel, to Pamela Klaber 'It was as important for the world of ceramics as any of the regular auction sales'. At its height, there were approximately forty-five to fifty dealers from all over the world, and everyone attended. Part of the fair's lure was its academic side, as people socialised and learnt from the series of lectures that were published in the following year's handbook. A magnet for all ceramic lovers – collectors, dealers and museums alike – the fair eventually petered out in the new millennium, but it undoubtedly advanced the world of ceramics in its time.

One of the grander post-war porcelain dealers was Albert Amor, which supplied porcelain ornaments, typically to adorn furniture from Mallett's or Partridge. Since 1943, the firm was based in Bury Street, a stone's throw from Christie's. The shop was a favourite of the Queen Mother, who built up a collection of Chelsea botanical wares. Albert Amor's guiding spirit was Mrs George, who arrived as a secretary in 1951, and by 1973 had taken over the business. She was described as 'a cross between the Queen and Mrs Thatcher who played the Royal Warrant very well'.[12] The shop had a reputation for stocking flamboyant pieces, including a good stock of Worcester porcelain, which was popular with collectors because of its availability, and appealed to American clients arriving in May.

In the context of English porcelain, Simon Spero recalls that, before the 1970s, you would rarely see a private client at auction, and it was the dealers who still clustered around the horseshoe. According to him, 1970–2010 was the golden period for collecting English porcelain; collectors from a broader group were drawn to the 'Blue and White' ceramics in which he specialised. For Spero, 'a fresh generation of dealers was emerging, and long-established traditions were becoming blurred as the distinction between dealer, collector and professional amateur was gradually eroded, eventually to the benefit of both'.[13] A publishing boom made the subject more accessible, as monographs, catalogues of collections and more authoritative auction catalogues appeared. Spero points out that, between 1965 and 1996, there were sixty-eight books written on English ceramics.

Bernard Watney, Geoffrey Godden, Henry Sandon and Spero himself all wrote books to provide what Spero called 'stimulating information'. Spero's *Price Guide to 18th Century English Porcelain* (1970), published by the influential Antique Collectors' Club, was the first handbook in porcelain values and part of the democratisation of the subject. He recalls that it was seen by some colleagues as a betrayal of the dealer's position, just as the appearance of the trade magazine, the *Antiques Trade Gazette* in 1971, was received 'with widespread alarm by the majority of the antiques trade, as a threat to their livelihood'.[14]

The rise of 'Blue and White' was stimulated by Bernard Watney's book, *English Blue and White Porcelain of the 18th*

Century (1963). Watney was a company doctor at the Guinness brewery, and a very knowledgeable yet potentially intimidating collector who had a special fondness for Liverpool porcelain, on which he wrote a standard work.* In response to the commonplace remark that something is rare, he would sometimes respond, 'Yes that is rare, I have three of them.' Watney observed that, 'A weekly visit to the sale rooms where it is possible to handle pieces and study them intelligently is more valuable than a whole library of books on ceramics.'[15] He would get up earlier than anybody else to catch the bargains at Portobello, and the great cry was always 'Has Bernard seen it?'

Spero was also central to the 'Blue and White' market, which has served him well, partly, he believes, because it chimed with the east coast American 'New England look'. He extended the interest in early English porcelain beyond Chelsea and Worcester to the factories of Bow, Limehouse, Liverpool and Lowestoft. A dealer with the instincts of an educator, Spero began young. At the age of twelve he was taken to Sotheby's, where he saw a Bow white porcelain fox *c.* 1750, which he could not afford (it sold for £240), but which inspired him to start collecting affordable cracked pieces. In a wonderful twist of fate, he found the fox thirty years later in a collection in Seattle. Spero spent a year at Knight Frank and Rutley's auction rooms, before opening a shop in Kentish Town in 1964 at the age of twenty. Anton Gabszewicz, the future head of Christie's English Porcelain Department, remembers his first visit to the shop: 'It was a minute room, no animal could have been swung there, the walls filled with laden shelves, a tiny counter with Simon apparently trapped behind it and, as far as I recall, nowhere to sit.'[16] Once Bernard Watney and fellow-collector Susi Sutherland turned up at the same time, jostling in this tiny space. They both wanted

* Simon Spero observed to the author that in the 1960s there was a preponderance of doctors who collected porcelain.

the same piece, a 'Blue and White' Derby coffee cup which, Spero recalls, 'Susi gripped in one hand, while perusing the shelves as Bernard looked on helplessly'.[17]

Spero moved to New Cavendish Street in 1972, dealing in early English 1750s porcelain, which has remained his favourite period. A decade later, he arrived in Kensington Church Street, where he opened with the exhibition *Early English Porcelain.* Spero describes this show as a disaster because he sold everything and had no stock left to deal in. He was saved by a call from Lady Reigate the next day, wanting to sell part of her collection. He recalls one of his more determined customers who so badly wanted a particularly rare early Worcester plate in one of his exhibitions in the early 1980s 'that he was impelled to queue outside my shop from 4.30 p.m. onwards on the afternoon prior to the opening day. Needless to say, he secured his just deserts.'[18]

Anton Gabszewicz had similar interests to Spero, focusing on the period 1745–55, the beginnings of English porcelain, and particularly Bow and Limehouse. His interest was encouraged by Father James Forbes, an Ampleforth monk, where Gabszewicz had spent a year in the novitiate. He joined Christie's under Anthony du Boulay in 1975, and six years later organised the Gilbert Bradley sale, exclusively of English 'Blue and White' porcelain, and an important marker in this taste.* In the same year, 1981, Sotheby's held the Louis Lipski sale, focused on Delftware. Lipski was co-author of *Dated English Delftware* (1984), and the sale brought a vast amount of material onto the market which appealed to the British collector Simon Sainsbury, as well as Americans who wanted the colonial Williamsburg look.

The most talismanic sale of the period – still often referred to – was of English porcelain and pottery removed from Rous

* Gabszewicz became head of Christie's European Porcelain Department in 1988.

Lench Court in Worcestershire, which exhibited a taste for the vernacular. The honours were divided between Sotheby's (1986) and Christie's (1990). Although the Rous Lench collection was quite academic, its strength lay in pottery which chimed with modernist tastes, Pimlico Road eclecticism and the burgeoning American market.

By the millennium, Sotheby's and Christie's property acceptance threshold of value had risen to a point where it became difficult to stage regular auctions of English porcelain and ceramics. Initially, Phillips and then Bonhams stepped into the breach as John Sandon and Fergus Gambon took over the mantle of Anton Gabszewicz, who left Christie's in 1992. John Sandon was the son of the celebrated Henry Sandon of *Antiques Roadshow* fame. Both were above all Worcester enthusiasts. To John would now fall the great sales, beginning with the three-part Bernard Watney Collection (1999–2000) while still at Phillips. This was the zenith of 'Blue and White', and the collection was especially strong on Liverpool and Limehouse. Sandon would also handle the famous Zorenesky Collection of Worcester porcelain, sold in three sales at Bonhams between 2004 and 2006. He produced definitive catalogues for the sales, demonstrating how far the market and knowledge had come. The post-war collectors of ornaments, once so numerous, had been replaced by specialists, experts and discriminating collectors who only wanted the best.

CHAPTER 13

Sculpture and Works of Art

'It is naught, it is naught, said the buyer. But when he is gone his way, he boasteth.' *

PROVERBS 20:14

The term 'works of art' was one the art world used to cover a multitude of areas: medieval, renaissance and baroque objects, marble and bronze sculpture of all periods up to the twentieth century, enamels, ivories, arms and armour, snuff boxes, Fabergé and, in the early days, even icons and violins. Sculpture was at the core of the category, and London dominated the auction supply. As the dealer Pat Wengraf argued, sculpture was an area where 'Dealers catalogues were at the cutting edge of scholarship and some way in advance of museums,'[1] What is certain is the close relationship between the museums and dealers, both in terms of the exchange of knowledge and building collections. The sale of sculpture (especially baroque) to American museums resembles and overlaps with the sale of *seicento* paintings as curators began to build their sculpture collections. At the Getty Museum, Peter Fusco had a brilliant run, from the de Vries *Dancing Faun* to his attempted acquisition of Canova's *Three Graces*.

The V&A was the powerhouse of scholarship for works of art

* Inscribed on an ivory plaque in Wartski's shop.

and sculpture. John Pope-Hennessy's three-volume *Catalogue of Italian Sculpture in the Victoria and Albert Museum* (1964) was the standard work, and as director of the museum he was supported by two knowledgeable deputies, Tony Radcliffe and Charles Avery. While 'The Pope' was remote and Olympian, Radcliffe and Avery were more willing to offer opinions to the trade. The crucial foundation of bronze scholarship was Wilhelm von Bode's three-volume *Italian Bronze Statuettes of the Renaissance* (1907–12), which reproduced all the most important Italian bronzes. Several new publications, notably Rupert Gunnis's *Dictionary of British Sculptors 1660–1851* (1953), led to more secure attributions in the areas they covered.

Before 1939, great medieval works of art had been much prized and fought over. The creation of New York's Metropolitan Museum's medieval outpost, the Met Cloisters, is the prime example of this interest. Medieval works were, however, in short supply after the war, and, as Sam Fogg commented, 'The 1960s and 1970s were not a good time to be buying medieval art, as there were virtually no sales for stock before the Robert von Hirsch sale in 1978.'[2] With fewer great medieval works of art for sale, baroque and neoclassical sculpture came into focus. Dealers pointed out the availability of high-quality works and how cheap sculpture was compared to paintings. Apart from the Getty, the leading institutional collectors in the field were the Metropolitan Museum and the Cleveland Museum (which had more funds than anybody else), followed by Minneapolis and the Clark Institute. The Getty made the early decision that it was going to be too difficult to make a medieval collection and concentrated on later periods – a policy they later regretted.

There were still medieval specialists operating in post-war London. Herbert Bier (one of the many brilliant émigré dealers who went from the Isle of Man to the Pioneer Corps) acted as the London buyer for the Cleveland Museum and for the

Canadian collector and media mogul Ken Thomson. John Hunt and his wife Putzel began dealing from St James's Place in 1933, offering medieval material to the Metropolitan Museum. Hunt gained the trust of that difficult, haggling customer William Burrell, and was to have a major influence on his collection. Living between Ireland and London during the 1950s, the Hunts formed a close friendship with Peter Wilson and John Hewett. Together they would go 'antiquing' in Portobello at the weekend, agreeing that Wilson would have first call on all bronze items they found and Hunt priority on all medieval ivories.[3] Wilson made him the Sotheby's consultant on works of art before 1500.

One of the most sensational pieces the Hunts sold was the Butler Bowden cope, a piece of English embroidery of the kind known as *Opus Anglicanum*. The cope was considered by the V&A to be the most important English vestment of the late thirteenth or early fourteenth century that was ever likely to come on the market. The Hunts sold it to the Metropolitan Museum in 1955 for the colossal sum of £33,000 but its export was stopped, and a special grant from the Treasury enabled the V&A to acquire it.

Thanks to Peter Wilson's interest in sculpture, Sotheby's had a stronger department than Christie's. As well as John Hunt, the team included John Hayward, Howard Ricketts and, later, Richard Camber and Elizabeth Wilson. Despite this expertise, John Mallet, a young cataloguer, remembers the dependency on outside knowledge: 'For difficult matters of authenticity in the hard-stone carvings or trinkets from the great Russian firm of Fabergé, we would call in Kenneth Snowman of Wartski's, whereas on several occasions I took disputable ivories to John Beckwith in the Victoria & Albert or Peter Lasko in the British Museum.'[4] At Sotheby's, the Stoclet sale in 1963 was a rare case of an illustrated single-owner collection with a delicious mix of Japanese wood carvings and medieval works of art.

One name is especially associated with the period's greater focus on baroque sculpture. Andrew Ciechanowiecki is probably the only London dealer to give his name to an anonymous master. The 'Ciechanowiecki Master' was responsible for a group of bronzes generally dated to the seventeenth century. Ciechanowiecki was almost unique in convincingly handling and knowing as much about sculpture as paintings and drawings. He was mentioned in Chapter 5 as a paintings dealer, but we must also consider him here in the context of sculpture, which was his main interest. At Mallett's, his first exhibition in 1962 was of works by the French *animaliers* – a term that Ciechanowiecki invented. He went on to stage exhibitions of terracotta sculpture by Jules Dalou and Jean-Baptiste Carpeaux, and by the time he established the Heim Gallery, sculpture was an equal component with painting. His greatest interest was in seventeenth- and eighteenth-century Florentine sculpture, and the dealer Danny Katz called him 'the man for the baroque', while Ciechanowiecki was also in the vanguard of promoting neoclassical sculpture.

Ciechanowiecki became the first sculpture dealer to produce serious scholarly catalogues, and John Pope-Hennessy claimed that he had 'changed the face of art dealing'.[5] He visited American museums every year and, as Pat Wengraf remembers, 'He told them what they wanted to buy, which was what he had.'[6] Prices were still low in the 1970s, but he got them into the habit of buying sculpture and persuaded them to place works by Soldani and Algardi underneath their baroque paintings. One of Ciechanowiecki's great coups was to identify the Soldani bronzes at Blenheim. In Rainer Zietz's opinion, Andrew educated Peter Fusco.

London was fortunate to attract brilliant foreign-born dealers like Ciechanowiecki. Rainer Zietz was the youngest of them. Born in Bad Lauterberg in the 1940s as the son of a physician, he studied medieval and early Italian renaissance art at Heidelberg University, where he attended seminars on European sculpture and decorative arts. Zietz began dealing in Venetian glass and art nouveau (a then little-understood field). He soon extended this: 'I realised that in certain areas important objects of museum standard were out of fashion and undervalued, for example Italian maiolica, Medici porcelain, Venetian glass, French Palissy and Saint-Porchaire ceramics, silver and sculpture. From the beginning of being a dealer in 1969, I never worked in one area.'[7] What encouraged Zietz to set up in London was a 1971 encounter with the collector Paul Wallraf and his wife Muriel, who introduced him to the world of international connoisseurs. Commuting between Hanover and a basement flat in Chester Square, it was in that basement that Zietz negotiated the greatest agreement of his life.

The Robert von Hirsch sale at Sotheby's in 1978 offered an unprecedented and unrepeatable opportunity for German museums to acquire some of the finest works of art that had ever come on the market. The collection ranged from Romanesque objects to works by Van Gogh (both paintings and drawings), although it was the medieval works of art that caused the greatest stir. The most important collection ever brought into London for sale from mainland Europe, the von Hirsch sale instigated federal cooperation between the German museums for the first time. Approached by Agnew's and Colnaghi, the museums chose to do much of their bidding through Zietz. It was a considerable matter for a young dealer to be entrusted with such a task. The negotiator at the German end was Herman Abs, the German banking doyen who, mediating between the various museums, came and thrashed out the details in Rainer's flat. Offering to bid for no commission because he keenly felt the honour of his position, Rainer received a small honorarium instead. One of the stars of the sale, the arm ornament of Frederick I Barbarossa *c.*1160–80, posed a problem as it came up early, making it difficult to know how to budget. Zietz secured the item for £1.1

million on behalf of the Germanisches Nationalmuseum in Nuremberg. Later in the auction, he bought the medieval enamel medallion representing *Charity*, one of the supreme achievements of Mosan metalwork, for £1.2 million for Berlin's Staatliche Museen. Another major buyer was the British Rail Pension Fund, which had entered the market a few years before, in 1974. The sale was a massive shot in the arm for works of art and the prices for important pieces rose accordingly. Zietz went on to sell items in his chosen field to nearly every major museum in the world.

The leading English-born dealer before Danny Katz got into his stride was Cyril Humphris, described by one colleague as 'secretive, bold, curmudgeonly and occasionally charming'.[8] As Zietz recalled, 'I saw him like a giant when I came to London.' Humphris left school at fifteen, joined the army at seventeen and then apprenticed with Alfred Spero, who operated from a shop in Knightsbridge full of sixteenth- and seventeenth-century bronzes and Hispano-Moresque plates where everybody could afford something. Charles Avery recalls going there: 'It was dark and covered with things under tables and everywhere, another world, mostly Italian maiolica and sixteenth-century bronzes.'[9] Such was his prestige that when Spero viewed an auction he would take care to give every bronze a minute's attention to throw competitors off the scent.

Humphris initially worked in partnership with an ex-actor, David Peel, selling maiolica and bronzes, before setting up his own Bond Street shop dealing in a thousand years of sculpture up to Rodin. He was a pioneer in promoting busts by eighteenth-century English sculptors such as Roubiliac and Rysbrack. 'When you buy works of art you have to be courageous,' Humphris once observed. 'It is a small market and expertise is everything.' Mad about horses, he was said to have brought something of the risk of the turf to the saleroom, and this was manifestly the case in his most famous purchase, the

Adriaen de Vries *Dancing Faun* auctioned by Sotheby's on 7 December 1989.

The highlight of Humphris's career was something of a drama and forms the most remarkable sculpture story during the period. The *Dancing Faun* was almost the 'sleeper' of the decade. Created in the Netherlands *c.*1615, the thirty-inch-high bronze statue had been bought unrecognised at Christie's in 1951 for £100 and placed in a rainy Sussex garden for almost forty years, where it acquired a beautiful patina. Hearing of a spate of garden ornament thefts in the area, the owner sent it to Sotheby's Billingshurst saleroom for their Garden Statuary sale with an estimate of £1,500/2,000. When the catalogue appeared, the Sotheby's London sculpture expert Elizabeth Wilson took it with her to lunch at the Chalet with the dealer Philip Astley-Jones. Having just seen the de Vries exhibition in Vienna, Wilson opened the catalogue while she was waiting and almost jumped out of her seat when she saw the small image of the *Dancing Faun*. One of the features of this bronze is that it was made from a single cast and is therefore unique. Elizabeth Wilson immediately had it brought up to London for thermoluminescence testing, confirming a date of *c.*1615.

When the *Dancing Faun* came up for auction in Bond Street with an estimate of £1/1.5 million, it caused a sensation. Humphris paid £6.82 million without a buyer for it. According to him, the client had backed out and he was left paying £5,000 a day in interest. As Humphris told a reporter, 'Suicide definitely was an option', but fortunately for him Peter Fusco of the Getty Museum stepped in and bought the sculpture. The sale captured the world's headlines and demonstrated that sculpture could command prices comparable to paintings.

In the same year as the von Hirsch sale, Christie's provided a sleeper. When Lord Lanesborough gave Christie's the contents of Swithland Hall to sell in 1978, the catalogue contained a bust of an unidentified pope that sold for £85 to Roy Pope, a suitably named Clapham antique dealer. Pope sold it for £240 to Nicholas Meinertzhagen, an antiquarian bookseller who identified it as Bernini's bust of Pope Gregory XV. He sold it at Sotheby's in 1980 for £120,000.* As a result of the Swithland Hall sale, Paul Whitfield, Christie's managing director, rang Charles Avery at the V&A: 'You may realise we are in a bit of a problem with this department.' With Avery's arrival at Christie's to run their sculpture department in 1980, Sotheby's faced real competition. Avery was knowledgeable, personable and something of a magnet. One of the areas he was able to develop was English school marble sculpture, and in 1985 he sold a Roubiliac of the 4th Earl of Chesterfield for £520,000. If it was a sleeper that brought Avery to Christie's, it was another that caused his departure, a sad story concerning an exceptionally knowledgeable and helpful expert.

It was in a Christie's Garden Statuary sale catalogue that Pat Wengraf spotted a half-length female figure catalogued as 'eighteenth century' which she suspected was a Giambologna. Ironically, Wengraf recognised it from the description in Charles Avery's own article 'Giambologna's "Bathsheba": an early marble statue rediscovered' in the June 1983 issue of *The Burlington Magazine*. The sale on 13 September 1989 carried an estimate of £3,000 to £4,000. She brought in Tim Bathurst of Artemis to be a partner in the purchase and decided not to attend the sale to maintain secrecy. Instead, she sent her husband, Alex, and their butler, Paddington, to do the bidding. Danny Katz had also spotted the Giambologna and asked Tim Clifford if he was interested on behalf of the National Gallery of Scotland, but he could only afford to go to £300,000. Paddington was given instructions to go on bidding until Alex Wengraf gave him the

* It would have a final saleroom appearance at Christie's New York in 1990, when it failed to reach the reserve price of $7 million. It is now in the Art Gallery of Ontario.

signal to stop. The bidding competition between Pat's butler and Danny's chauffeur was won by the former, paying £715,000 and Wengraf later sold it privately. Astonishingly, Charles Avery himself had left a low bid, a mistake that cost him his job.

A natural businesswoman, and inspired by the *Giambologna, Sculptor to the Medici* exhibition at the V&A in 1978, Pat Wengraf had turned her interest in sculpture into her business. Entirely self-taught, she bought her first piece of sculpture for trading in 1979, a terracotta figure of St Michael now in the Metropolitan Museum. John Pope-Hennessy gave Wengraf her first big break the following year when he introduced her to Jayne Wrightsman, who bought four bronzes, although Wengraf is not certain that she really liked bronzes. Pope-Hennessy also introduced her to the Argentinian collector Claudia Quentin, who became a great client and friend. Wengraf operated initially by appointment out of Bury Street and then moved to Jermyn Street above Andrew Ciechanowiecki, whom she found very astute but controlling. A forceful personality (she once took the French Museums, the Ministry of Culture and the French Customs to court and won her case), she felt that Cyril Humphris, Alain Moatti (in Paris) and Rainer Zietz had too much control over Peter Fusco. Refusing to be overlooked, she went to John Walsh, Fusco's boss at the Getty, to complain and thereafter got some deals. She concentrated on renaissance and baroque sculptures because of their availability and because she found much neoclassical sculpture a little cold. As Iona Bonham-Carter, who had worked at both Sotheby's and Christie's, pointed out, 'Nobody was then interested in marble neoclassical busts except Christopher Gibbs until the *World of Interiors*.'[10] The art historian Hugh Honour's groundbreaking exhibition *The Age of Neo-classicism*, held in 1972 at both the Royal Academy and the V&A, was pivotal in changing attitudes. Canova, the hero of the exhibition, was now seen as very desirable.

The apogee of the taste for neoclassicism would be the

drama over Canova's *The Three Graces* sold to the Getty in 1989. The sculpture was of unusual interest, commissioned in 1814 by the 6th Duke of Bedford, who created a *tempietto* for it at the end of his sculpture gallery at Woburn. First offered under the Acceptance-in-Lieu (of tax) scheme to the Fitzwilliam Museum, the museum's preference was for it to remain *in situ.* This was allowed under the scheme if there was public access, but the family wanted to use the space for retail purposes. *The Three Graces* was sent to the *Treasure Houses of Britain* exhibition in Washington (1985–6), and at some point its ownership was transferred to a Cayman Islands trust. When a sale to the Getty was announced, in the storm that followed it was pointed out that the sculpture had been exported only on a temporary licence for the exhibition. The saga dragged on until 1994, when *The Three Graces* was acquired jointly by the National Galleries of Scotland and the V&A for £7.6 million, a price tag which showed how far the Canova market had come.*

One London dealer revived interest in the Middle Ages by approaching it from a different angle. Sam Fogg trained at the Courtauld and started his commercial career at Portobello, selling Penguin books and art reference volumes. When he moved into medieval manuscripts around 1986, he was by far the youngest dealer at every sale. The book dealers approached manuscripts in the manner of bibliophiles, whereas he approached them as works of art, selling three-dimensional objects alongside them. There was not much competition in London, as most of the dealers, notably Jacques Kugel and Alain Moatti, were in Paris. 'Nobody was doing medieval in London at the time,' recalls Fogg, 'the exception being Rainer Zietz, who has a Continental sensibility.'[11]

Fogg found that buyers of contemporary art liked placing

* The previous record for a Canova was a life-size marble *Dancer*, bought from Colnaghi by the Berlin Museum in 1981 for about £600,000.

it alongside cult objects of medieval art in an extension of the Breton/Malraux taste. He learned how to exhibit medieval pieces from contemporary art dealers who were ahead in the way they thought about presentation. Fogg started doing six exhibitions a year, producing scholarly catalogues in the manner of his American hero, the book dealer Hans Kraus. He generally went up to about 1520, or a century later with Northern art. One collector describes Fogg as the master of the oblique sale, offering a tapestry to a manuscript collector and a manuscript to a sculpture collector. He was fortunate with his clients, who included the London businessman Paul Ruddock and the transatlantic collectors Ken Thomson and Ronald Lauder. Hermann Baer (not to be confused with Herbert Bier) had originally introduced Thomson to Kunstkammer material and ivories. The high point of Fogg's career was buying two of the ex-von Hirsch treasures when they were resold at the British Rail Pension Fund sale at Sotheby's in 1996. These were the Romanesque bronze *Base for a Candlestick*, related to the Gloucester Candlestick in the V&A, and the Becket *Chasse*, the latter acquired for Thomson, although the item was stopped at export and went to the V&A.

For the last forty years, one individual has dominated the London sculpture market: the irrepressible Danny Katz. With a zest for life and boyish enthusiasm, Danny was described by one colleague as 'a Peter Pan who gets away with things'. His story is a remarkable love affair with sculpture and a triumph over adversity. The Katz family had an antique shop in Brighton, where Danny got his first teenage taste of the trade. For him, book learning was impossible, as he suffered with the neurological condition Tourette's. At twenty he moved to London and worked for his uncle and first mentor Cecil Lewis. He bought bronzes for £5 and sold them for £8 irrespective of their date. It was museums above all that gave Katz an education. At the age of eighteen, he entered the V&A for the first time and saw Giambologna's *Samson slaying the Philistine*; as he recalled, 'I knew then that my life's work would be spent in sculpture.'[12] The collector Michael Travers was the first to recognise that the young Katz had an extraordinary visual talent, introducing him to John Pope-Hennessy, who took no interest, but his colleague Tony Radcliffe became Danny's second mentor.

Katz was a groundbreaking dealer by virtue of the risks he was prepared to take. As a sculpture dealer in London operating around the renaissance and baroque, he followed the pioneering work of Andrew Ciechanowiecki and Cyril Humphris. He would assiduously take things to the V&A to be examined by Radcliffe and Avery. Just as they were beginning to find this tiresome, Katz appeared with an image of a ravishing marble by Giambologna. Following Danny's discovery of the artist's *Bathsheba* at a Swedish country house in 1982, Charles Avery flew out with him to inspect it. The subsequent sale to the Getty gave Katz the financial comfort that enabled him to start setting things aside.

Katz sold mostly to museums, as it was hard to find a private customer for sculpture at the top level. As he pointed out, 'A client normally lasts about seven years – this could either be the attention span of a collector or it could be the period in office of a curator.'[13] The Kimbell Art Museum of Fort Worth, Texas, was a big client, to which he sold a German *Virgin and Child* (1486) made in silver and set with precious stones, for approximately $10 million. This late Gothic work at first seemed too good to be true. One client who was halfway between being private and public was the Prince of Liechtenstein, to whom Katz sold one of the most beautiful bronzes: Jacopo Sansovino's *Saint John the Baptist* of *c.*1540 from the Alfred Beit collection.

Rising prices brought out a great deal of material during the 1990s, particularly in France, where the market was opening up. As Katz reflected, 'You needed to make the clients realise that if you could buy a Monet costing $40 million, a Canova that costs only $2 million is something of a bargain.'[14] By the new millennium, his extraordinary personal collection of Modern British paintings and old master paintings was absorbing much of his energy. He turned largely to philanthropy and gave substantial sums of money to the museums that nurtured his interest, particularly the National Gallery, the Fitzwilliam Museum and the Ashmolean.

Chapter 14

All that Glisters: Silver

'If the electric light started the process of making silver less desirable, the advent of the kitchen dining room meant that the middle class no longer wanted table silver.'[1]

Timothy Schroder

Pick up a copy of any antiques magazine from the early 1970s and between the mahogany sidetables and sporting paintings are advertisements for London silver dealers, now mostly disappeared: Partridge, Tessier, Michael Welby, How of Edinburgh, Simon Kaye, and some survivors, S. J. Phillips, Koopman and S. J. Shrubsole.* They evoke a time when people still bought table silver for dining rooms. Timothy Schroder put it succinctly in suggesting that electric light and eating in the kitchen eroded this need. As he explained to the author, 'Silver, when illuminated by flickering candlelight, comes alive and almost dances before the eyes, but when lit by electric light it becomes flat and dead.'[2] Domestic and economic changes may have worked against the market, but the London silver trade remained buoyant, thanks to the competition of collectors seeking grand display silver at the top end, and the buyers

* S. J. Shrubsole survives only in New York.

of 'collectables', like spoons and wine labels and 'novelties', at the bottom. Historically, silver has been, and still is, an important element in the business of 'show' visible in private houses, churches, government and diplomacy. Another factor that came into play was the systematic collection building of certain American museums over the period. Boston, Huntington Art Gallery and Williamsburg, among others, were largely supplied by London dealers.

Silver offers many faces: for some it represents wealth, art and history, and for others it is also a commodity for its base value, like gold; but, as every dealer will tell you, there is no correlation between the value of silversmiths at the highest level and that of the metal. Changes of taste were significant in impacting value, as Regency silver and the work of Rundell, Bridge & Rundell, the Royal Goldsmiths and their manufacturing partner, Paul Storr, became as sought after as the ever-desirable early eighteenth-century Huguenot makers. Even more striking is the change in attitude during the 1970s towards Victorian commemorative silver, which shifted from its melt value to being enjoyed as sculpture. John Culme, a former director of Sotheby's and historian of the silver market, noted that, 'While the trade was largely content to carry on their businesses in much the same way as had their pre-World War II predecessors, there was an increasing interest in the history of silver and the lives and times of its makers. Probably the most influential individual in this, mixing an academic approach with a keen commercial awareness, was Christie's silver director, Arthur Grimwade.'[3] In the view of Lewis Smith of Koopman, the auctioneers controlled the market during the period of this book, and none more so than Grimwade.

Arthur Grimwade (1913–2002) was an exception to the ranks of Etonian directors at Christie's. Head of the silver department from 1954 to 1979, the undisputed leader of his profession, he was also a first-class, self-taught academic. Perhaps Grimwade's greatest legacy is *London Goldsmiths, Their Marks and Lives* (1976), a consummate work of scholarship in which he identifies the marks of hundreds of eighteenth-century silversmiths, alongside a biographical dictionary of 2,600 of them. One of his colleagues, Christopher Wood, described him: 'tall, bony and irascible, Arthur had a face gadrooned like a piece of old English silver. His empire was the Strong Room, in the basement, where one could occasionally hear him hurling some wrongly marked piece of silver against the wall.'[4] Grimwade was old-fashioned, and during the late 1960s he attempted to ban female employees from wearing either boots or trousers. According to Wood, 'most of the girls reacted by wearing the shortest possible mini skirts'.[5] Grimwade wrote a delightful diary, *Silver for Sale* (1994), which is the only published account of pre-war Christie's.

Mockingly referring to his beloved silver as 'tin', Grimwade had one immeasurable advantage over all his competitors. In the 1930s, he had started compiling a card index that listed every important piece of silver that Christie's had ever sold, back to 1830. This hundred-year record meant that he could reconstruct the entire history of most pieces he was offered. Grimwade was responsible for huge numbers of silver sale catalogues, establishing standards for auction-room cataloguing and remaining indispensable to students. In the years immediately following the war, silver flooded onto the market, and Christie's held more sales than at any other time since. During the 1960s and 1970s, Sotheby's, Christie's and Phillips were each selling 200 to 250 lots a week. Fortunately, given the quantity, Grimwade was also a superb auctioneer.

Sotheby's, like the other auction houses, benefited from the increased activity in the market for antique silver, but was at a disadvantage. Head of its Silver and Jewellery Department was Fred Rose, described by Culme as 'a Lincolnshire auctioneer's clerk who had joined Sotheby's in 1936. Although his first interest was in furniture, he was obliged to turn his attention to other areas of the market during the war.'[6] He was joined in 1956 by Richard Came, whose strengths as head of the department after Rose's retirement in 1965 lay in dealing with clients and securing property for sale. A flamboyant personality, Came would ride his 'bone-shaker' bike to the office at break-neck speed, wearing a pinstripe suit and bowler hat. Unlike Grimwade, Came was not academically minded, but he was Sotheby's most talented auctioneer after Peter Wilson. Given Grimwade's ascendancy, Sotheby's were fortunate to be given the landmark sale of the Berkeley Castle Dinner Service in 1960. In this world record-breaking sale, a Louis XV silver service, consisting of 168 pieces made in Paris by Jacques Roettiers 1735–8, was bought by Partridge on behalf of Stavros Niarchos for the colossal sum of £207,000.

Most of the major post-war sales were held at Christie's. When the Earl of Lonsdale abandoned Lowther Castle he sold much of the family plate there in 1947 for very modest prices. The Lonsdale version of Rundell's Shield of Achilles designed and modelled by John Flaxman (one of five originally made in 1821 and 1822, the first of which was purchased by George IV), made only £520.* By the time Christie's sold Lord Brownlow's silver, prices were rising, particularly for rarities. This can be followed through the large pair of James II tankards with Chinoiserie decoration made in London in 1686, which had descended through the Brownlow family. First sold at Christie's in 1963, they realised £17,000; they reappeared in the saleroom in 1968, selling for £56,000, and then again at Sotheby's New York, in Jaime Ortiz-Patino's sale in 1992, when they sold for $797,500.

* It was acquired by Huttleston Broughton, 1st Lord Fairhaven, and bequeathed to the National Trust along with Anglesey Abbey in 1966.

John Culme, who began his career at Sotheby's in November 1964 as a bright eighteen-year-old dogsbody in the silver department, provides an evocative picture of the auction room in those days:

That summer the firm had acquired Parke-Bernet and during my interview, building work was underway to join the recently purchased 4/5 St. George Street with the back of 34/35 New Bond Street. Sotheby's was expanding as I arrived and to the Silver Department were added two new employees: Kevin Tierney and, a month later, myself. Kevin and I were both on steep learning curves because neither of us had any knowledge of silver or hallmarks, &c. Kevin was the junior cataloguer and I was junior porter to the combined silver and jewellery departments under the then director, Fred Rose. In fact, my duties were to assist Macleod, our porter (who spent most afternoons in The Masons Arms in Maddox Street) in numbering sales of jewellery, silver, objects of vertu and watches as well as running errands (to Goldsmiths' Hall, the gem testing laboratory in Hatton Garden, &c., &c.), cleaning silver and any other odd jobs that needed attention.[7]

As Culme recalled:

Kevin and I were quick learners. No wonder as at that time during the season we had three sales per month (on Thursdays, the fourth week was for our jewellery sales), with about 250 or 300 lots in every sale. This schedule was matched by Christie's where sales of silver and jewellery were on Wednesdays. Phillips, Son & Neal had regular silver sales, too, so there was plenty for the London trade to get its teeth into. In those days our per lot threshold was raised to £75, so as you may imagine all these sales were full of everything from soup tureens and epergnes (in our better sales) to snuff boxes and wine labels. Good examples of seventeenth-century and earlier English silver were much in demand, as were Queen Anne Huguenot pieces. Trade buyers most in evidence for these were Hugh Jessop, Tom Lumley (who had worked for Sotheby's in his youth in the 1930s), S. J. Phillips and the terrifying 'Ben' How of Pickering Place, widow of Commander How, dealers in all the best items, with an enviable list of private customers whose names were unknown to us. Commander and Mrs. How were also the leading experts on old spoons, which are important to specialists because of the information they hold about early silversmithing throughout the country. A large number of the remaining regular trade buyers at auction were based in London: Simon Kaye Ltd of Albemarle Street, Bloomstein, Swonnell and others of the Bond Street Silver Galleries, John Bourdon-Smith, C. J. Vander (the old established manufacturing silversmith, whose antiques department was energetically run by Richard Vanderpump), and many more. Simon Kaye was probably the heaviest buyer, with a large network of smaller dealers across the country on his books as buyers. Nearing retirement, his authority as leader among trade buyers was challenged at a sale at Sotheby's in the late 1960s when Jacques Koopman, a relative newcomer, bid on and secured almost every lot. From then on it was Koopman, not Kaye, who was the heaviest buyer.[8]

In terms of taste in silver, the rise of Regency is the most salient trend, and the reappraisal of the most productive firm, Rundell, Bridge & Rundell, whose work is so often stamped with the mark of Paul Storr. Alastair Dickenson recalls that there were still people around who had begun their careers in the 1920s and who thought of Paul Storr as 'modern silver'.[9] The silversmith was the subject of N. M. Penzer's seminal book, *Paul Storr, Last of the Goldsmiths* (1954), which sparked a frenzy among a group of wealthy collectors.* The problem with books of this sort is that they encouraged collectors to think of silversmiths as individuals: 'names' to latch onto, as with Rubens or Gainsborough, which was misleading.

The effect on the market can be judged by the Christie's sales from the collection of the late Mary, Princess Royal and Countess of Harewood in 1965 and 1968. They included a group of silver-gilt banqueting plate with the maker's mark of Paul Storr. Christie's illustrated the silver gilt in colour to stunning effect. Amongst the buyers was Sir William (Billy) Butlin, of holiday camps fame, who enjoyed the brio of the Regency and built up an important collection. Butlin would have his own sale at Christie's (17 July 1968), although his flamboyant taste was still at odds with many of the dealers and collectors, 'who maintained that nothing of interest or worthy of attention was made after about 1800'.[10]

The dealer who would pick up and run with Regency taste was Jacques Koopman (1931–1991), who recognised its great potential. He was to inspire a new generation of silver collectors to appreciate its beauty. When Sotheby's sold a collection of silver and silver-gilt belonging to the Duke of Northumberland in 1984, the most important piece was the one of the five Shields of Achilles acquired by the 3rd Duke.† In contrast to the 1947 Lonsdale price of £520, this one fetched £484,000. Against stiff competition, Koopman bought the Northumberland Shield for stock, marking him out as the leading silver dealer. The sale also signalled the point when the nineteenth century had caught up with the eighteenth century. Koopman later sold the shield to Mahdi Al Tajir, one of the two major collectors of the period,

* Paul Storr would continue to attract academic attention over the next twenty years, especially during his time as partner of Rundell, Bridge & Rundell.

† The shield was designed and modelled by Flaxman, mark of Philip Rundell for Rundell, Bridge & Rundell, London, 1821.

the other being the London-born Californian real estate developer Arthur Gilbert, who was one of the underbidders.

Koopman was generally described as nervously active, impatient, constantly on the move, with no time for small talk. A strong personality, he kept his Dutch accent. When he was a teenager his parents died in Auschwitz, but he had managed to jump off the train and make his way back to Amsterdam. One dealer remembers that 'Koopman was very tough, courageous, and a game changer by the prices he was prepared to pay'.[11] The business was founded in Manchester by his older brother Eddy Koopman, but from 1959, Jacques operated from the front corridor of the London Silver Vaults (No. 25) at Chancery Lane. Lewis Smith commented that 'having a vault on the front corridor was much like having a qualification of being a lawyer', such was the prestige of the position. The vaults, where fifty or sixty dealers had premises, were the epicentre of the silver trade, and began to attract retail custom during the 1970s. Koopman also exhibited at fairs in New York and Maastricht, where he found international clients, meaning that he didn't need a West End shop.

Koopman was responsible for at least two major silver collections associated with London: the most famous being that of Mahdi Al Tajir, a notoriously late payer. Al Tajir was, however, princely in his purchases and gave the market a welcome boost just as silver was going out of fashion. When Christie's exhibited his collection in 1990, it demonstrated the move of buying away from table silver to big display pieces. Another major collector whom Koopman advised was the Australian tycoon, Kerry Packer. 'As to Koopman's importance,' John Culme told the author, 'I would say that by his energy and nose for publicity as well as his keen understanding of business, he had a profound effect on the silver trade.'[12] After Koopman's death in 1991, Lewis Smith took over the firm.

If Koopman served the new collectors, one exceptionally

well-connected dealer amongst old money was Jane How (1915–2004), always known as 'Ben' to her friends (from her maiden name, Benson), but more often, 'Mrs How'. Diminutive, with a deep voice, and a law unto herself, her eccentricities and gruff manner have become part of trade folklore. Mrs How's interests were early English, Scottish and Irish silver from the thirteenth to the seventeenth century, and above all spoons. A neat figure in her tweed twinsets, she was accompanied everywhere by her huge, drooling English mastiffs (which by all accounts stank), whether she was walking them around St James's or determinedly driving them in her Bentley Turbo.

A long-standing member of the Antique Plate Committee at Goldsmiths' Hall, she met her husband, Commander George Evelyn Paget How, when they were cataloguing the Ellis spoon collection for Sotheby's in 1935.* Together, they built up a formidable dealership, confusingly known as How of Edinburgh, although it was located at Pickering Place behind Berry Brothers in St James's. Mrs How is, however, mostly remembered during her long widowhood. Helping to form many private and public collections in America, she focused on early English silver up to the middle of the eighteenth century and could claim the Aga Khan as a client. She trained many young dealers, including Hugh Jessop, John Bourdon-Smith and Brand Inglis, all of whom dined out on her crusty encounters with customers. After her death, Woolley & Wallis in Salisbury sold her personal collection, which attracted all the major spoon buyers of the world. The sale was indicative of the rise of provincial salerooms, as Sotheby's and Christie's focused on other markets.

Interest in the 'middle market' of domestic silver was maintained by the professional classes who had spare cash and who acquired useful items such as candlesticks or sauceboats. Like

* Their monument is the three vast folio volumes of *English and Scottish Silver Spoons and Pre-Elizabethan Hallmarks on English Plate* (1952–7).

brown furniture, these are now less highly regarded and prices have plummeted. Middle market buyers were also interested in 'collectables': spoons, 'smallwork' comprising novelties (snuff boxes, vesta cases, etc.) and wine labels, sometimes referred to as bottle tickets. During the late 1960s, prices were often extraordinarily high for such small items. The Bond Street Silver Galleries, with over twenty dealers spread over three floors, were much frequented by professional people during their lunch hour.

The greatest revolution in taste was the rehabilitation of Victorian silver, which, until the 1970s, was still being melted down. As so often happens, the change was preceded by scholarly interest, notably the publication of Patricia Wardle's *Victorian Silver and Silver Plate* (1963) and Shirley Bury's impactful series of three articles, 'The lengthening shadow of Rundell's', published in *The Connoisseur* in 1966. It was two young cataloguers at Sotheby's, John Culme and Bruce Cratsley, who saw the commercial possibilities, and were to organise the first sale of nineteenth-century silver against the staunch opposition of their boss, Richard Came.

John Culme recalls:

> Bruce (enthusiastically supported by me) thought that a specialist sale devoted to silver from the reigns of William IV and Queen Victoria would be a good idea, not only to encourage what we perceived as an emerging market but also to allow us to learn more about the subject which until then had been ignored. Richard was furious and balked at the idea, but he surprised us one day by agreeing to the sale. So, the first of a series of such sales was held at Sotheby's on 27 March 1969. It was a success. What I didn't know at the time was that it was Peter Wilson [Sotheby's chairman] who had approved of the sale because he was already making plans to open Sotheby's Belgravia, the first auction house in the world specialising in

> post-1830 pictures and works of art. Our first sales were held there in September 1971. The average rate of unsold lots in the Silver and Objects of Vertu Department at Belgravia between 1971 and its closure ten years later was never more than about 5 per cent.[13]

The benchmark sale for Victorian commemorative silver was the Mentmore Towers disposal in 1977, where elaborate and hitherto unsaleable sculptural trophies made unexpectedly high prices. In 1982, when the speculator Nelson Bunker-Hunt attempted to corner the silver bullion market, pushing the price to £20 an ounce, a new high for silver, Sotheby's put out an advertisement: 'Don't melt in the heat of the moment', referring to the way in which most Victorian silver had been treated until that time.

One dealer who was part of the new vogue for Victorian silver was Tom Lumley, who had broad and fascinating tastes. An elegant figure with an exquisite eye, he bought the best English silver of all periods, including things that were not in fashion: Victorian silver, aesthetic movement silver and Central American colonial silver. Lumley loved anything that, in his own words, was 'dotty', and he was influential in the formation of American museum collections at Williamsburg, Chicago and Boston. At the other end of the scale to Victorian, he had a taste for plain English silver of the early eighteenth century, which he sold to Victor Rothschild and Tommy Sopwith (of Sopwith Camel fame). Lumley operated from Bury Street 1937–65, then set up the Partridge's silver department before coming to rest on the third floor of No. 2 Old Bond Street. Like Mrs How, he drove a Bentley and always wore white gloves when inspecting silver.

Dominating Bond Street was the powerful and magnetic S. J. Phillips. Facing Sotheby's front door, the shop attracted customers with the delectable jewellery in the window, seducing them into buying silver. The furniture dealer Martin Levy learned a lesson 'from one of the strongest dealerships in London, S. J. Phillips. They have, amongst their gorgeous multi-million-pound jewels and works of art, items such as silver jar lids for pots of Marmite or jam, which are reasonably priced. Nobody needs to walk into S. J. Phillips without buying something.'[14] The shop had great appeal for grandees, as it managed to convey an old-fashioned, over-the-counter elegance. Martin Norton's two sons served in the front-of-shop while in the back was the wise, shrewd old man, who saw and knew everything, and was always tough on a deal.

Founded in the mid-nineteenth century, S. J. Phillips had been located in New Bond Street since 1873, first at No. 113 and from 1966 in No. 139, where the Victorian look of No. 113 was recreated. From its early days, it was a toy shop for the rich, or what Arthur Grimwade described as 'the greatest treasure house of acquirable possessions for limitless pockets of the world'.[15] When the military deceivers of 'Operation Mincemeat', better known as 'The Man Who Never Was', wanted to add authenticity to their fictional British officer washed up on the coast of Spain, they stuffed an invoice into his pocket from S. J. Phillips for a £53 engagement ring.

Queen Mary was a customer, and Mrs Thatcher had once gone there to buy a hairbrush for the bald Mikhail Gorbachev but bought herself a brooch instead. Their best-known client was the Californian collector Arthur Gilbert, who initially focused on 'the two Pauls' (de Lamerie and Storr). As time went on and opportunities for meaningful additions in these areas became more limited, his range of interest widened to earlier periods of English silver and to Continental, especially German, gold and silver. He ended up creating one of the first true *Schatzkammer* (treasure chamber) collections since the early twentieth century.

One day in early 1996, Lord Rothschild received an unusual

telephone call. It was from Nicholas Norton of S. J. Phillips: could he drop everything and come round to the shop to discuss a matter of national interest? It is not every dealer who can make such a summons. What emerged was the story of Arthur Gilbert and his search for a home for his vast collection of silver. Lord Rothschild was engrossed and spent the next three years landing this great whale, first on the banks of the River Thames at Somerset House, and latterly at the V&A, where it can be seen today.

There were many other distinguished dealers in London, notably S. J. Shrubsole, which had a New York branch selling top-tier silver. Two shops which cannot be overlooked are Garrard and Asprey. Garrard, the Crown Jewellers, founded by George Wickes in the reign of George II, kept up the tradition of manufacturing goldsmiths and jewellers, while maintaining an antique and second-hand silver side. Selling silver was a recent innovation for Asprey's, dating from when John Asprey invited Alastair Dickenson to develop the department. The firm had stellar clients, including the Sultans of Oman and Brunei. John Asprey gained these clients by going to their countries, and being patient and persistent. As a result, buyers from the Middle East dominated their balance sheet. Dickenson recalls John Asprey one day outbidding Mrs How on an important Queen Anne toilet service at Christie's. Paying £172,800, he was roundly rebuked by her: 'Why don't you leave buying important silver to the dealers and stick to retailing.'[16]

In 2003, the *Antiques Trade Gazette* announced the closure of the Bond Street Silver Galleries, which, along with the London Silver Vaults in Chancery Lane, had been the regular stop for silver and jewellery buyers when visiting London. By the millennium, the trade in antique silver was a shadow of its former self, and the number of dealers today is miniscule compared to the 1960s. Auctions have largely moved to provincial salerooms, with high-value items accepted by London houses placed in multidisciplinary sales to appeal to a wealthy international clientele. On the positive side, the arrival of Koopman under Lewis Smith in Dover Street means that silver can still be bought and enjoyed a stone's throw from Bond Street.

Chapter 15

Art Commodified: British Rail Pension Fund

'Buying pictures for love of art is virtually a thing of the past.'[1]

Robert Wraight in 1965

In the early 1960s, the notion of art as investment began to take hold of the market. The Goldschmidt sale had shown the power of uniting European and American collectors and dealers at international sales in London. In *The Economics of Taste* (1961), Gerald Reitlinger observed that modern picture sales and breaking records had become synonymous. He wrote that, 'The notion of Art as an investment has created more press publicity for auction sales than has ever existed before.'[2] The American banker Richard H. Rush's *Art as an Investment* (1961) underlined the new interest in art as an asset class. In 1966, Peter Wilson told the BBC *Money Programme* that, 'Works of art have proved to be the best investment, better than the majority of stocks and shares in the last 30 years.'[3]

Peter Wilson thought that publishing charts showing the steep price rises in impressionist art since 1950 would tempt investors. Stanley Clark, the Sotheby's publicist, and the business editor of *The Times*, had lunch in 1967 and dreamt up a Dow Jones index of the art world. Clark had always wanted to take the saleroom reports away from their low billing between the crossword and bridge columns, and place them in home news. He saw two ways to achieve this: headline record prices and good stories. Publishing a chart of rising values would be achieved thanks to the efforts of Geraldine Norman (née Keen), *The Times* statistician, an Oxford mathematics graduate who compiled the Times-Sotheby's Index. After five or six weeks working on it, she told the author, 'I decided it couldn't be done.'[4] She had, however, underestimated the determination of Peter Wilson. Norman described the method she finally adopted:

> On Wilson's advice I started with impressionists. Sotheby's kept their art market archive on cards, covering both Sotheby's and Christie's sales. Each card had a photograph of a painting, when it was up for auction, whether it was sold and the price, stretching back over the period we were interested in, that is since 1950. I devised a method, choosing six individual artists and getting Michel Strauss, the head of the Impressionist Department, to arrange each of the cards in categories ranging from good to better to best. It was thus possible to match paintings as they were sold with their appropriate category. The index was based on the movement of each category, averaged out for the artist – the artist's index. Then the six artists were combined into an overall impressionist index.[5]

Norman used roughly the same method for eleven other fields, including silver, Chinese ceramics, English pictures, and old and modern books. A problem arose with old master paintings when the head of department, Carmen Gronau, dug her heels in and said it was impossible to index old masters. Wilson, however, was adamant: 'You've got to do it, Carmen … you and Geraldine must go at it this evening. I want an index by the morning.' Giving them a bottle of whisky, he shut them in the department for the night. Geraldine recalls: 'We emerged tipsy with the makings of an index completed.'[6] The variation for old masters involved putting a contemporary valuation on each card and compiling the index based on seven schools rather than artists, and then averaging the price movements. Scientific it was not.

The first index appeared in the review section of *The Times* in November 1967. Peter Chance, chairman of Christie's, was incensed by the whole idea and took an instant dislike to Geraldine Norman. The index revealed that the value of old master drawings had increased by twenty-two times between 1951 and 1969, while for old master prints the multiple was thirty-eight times and Rembrandt prints forty times.[7] The publication provoked a storm of reaction, that Sotheby's were debasing artists' creative achievements, and whilst some dealers shared this view, others took advantage of it. *The Observer* suggested that the radio programme the *Stock Market Report* should round off prices with news from the art market: 'Renoir, steady; Vlaminck and Manet, mixed.'[8] Despite this derision, Norman remembers: '*The Times* knew that it had got a scoop and syndicated publication rights to the *New York Times*, the *Süddeutsche Zeitung* and *Connaissance des Arts*. So, the first index of art prices – and Wilson's chart wriggling upwards – got an international audience.'[9] As Philip Hook observed, 'From now on art was even more closely held in the guilty embrace of money.'[10]

Norman's charts appeared with a commentary and, in her own words, she 'gradually began to understand quite a lot about the auction market'.[11] In 1971, the silver market was collapsing, and Wilson closed the index, as he didn't like graphs that went down. Equally important, Norman had been appointed saleroom correspondent for *The Times*.[12] From now on, her relationship with Wilson and Sotheby's would be tricky. She rejected the ideas behind the index when she wrote a feature in 1974: 'The time has come to dismember once and for all the idea that art is a safe investment medium … the idea that art is a solid and safe investment medium is a fallacy.'[13] That contention would be tested by the first art investment fund to be established with scientifically measurable results.

The British Rail Pension Fund made its surprising appearance in the art market in 1974. It was significantly different from all previous investment funds, with access to far greater resources, much wider expertise and, most important of all, no time pressure. The fund was the brainchild of Christopher Lewin, one of the people responsible for the annual investment of the railway-workers' pension fund. As the 1970s progressed, inflation rose to over 15 per cent, real estate had lost some of its value (there would be a crash in 1974) and equities had stagnated. The fund was looking to diversify as a hedge against inflation. Lewin was himself a book collector and had observed that antiquarian books were outpacing inflation. Influenced by the writings of Gerald Reitlinger, he began to conceive the idea of a small percentage of the available funds to be put into works of art: 'The fact that prices had dipped encouraged me to come into the art market.'[14] The Times-Sotheby's Index had created a statistical justification for such a fund.

Christopher Lewin approached Peter Wilson. To Sotheby's, the idea was a godsend, providing a major new buyer at the top end of the market when it was most needed during a downturn. In times to come, the fund would also provide a series of superb sales. Lewin was impressed by Peter Wilson: 'I thought him the master showman.' 'The big question in my mind,' Lewin told the author, 'was whether some of the experts might channel the wrong works of art through us.'[15] Wilson had the answer, which was to have the fund and its purchasing managed by someone independent. He chose Annamaria Edelstein, who had worked at Sotheby's Publications and was the editor of their annual review, *Art at Auction*. Highly intelligent and strong-minded, she was able to keep Sotheby's in line. As she later put it, 'Peter Wilson also knew that I was bloody-minded and thought that was a good characteristic for somebody who had to put a collection together without being influenced by anyone, including him.'[16]

It was announced just before Christmas 1974 that the British Rail Pension Fund had decided to invest 3 per cent of their capital in works of art, relying mainly on the advice of Sotheby's. An independent company, Lexbourne, was set up to handle acquisitions. Edelstein was more independently minded than people outside imagined and would ask specialist departments to make suggestions. She allocated budgets to each area and set up a panel of department experts, but would also take advice from the trade and Christie's. Edelstein recalls: 'I never had so many friends.'[17] Sotheby's had to guarantee the authenticity of the recommended objects and their provenance, as well as the authenticity of items from dealers, for which they received a 5 per cent purchase premium from the fund. The amount the pension fund proposed to bid in an auction sale was kept secret from Sotheby's, and the bids were executed by dealers (often from different specialisations) so that Sotheby's would not know when British Rail were bidding.

Edelstein occasionally had difficult moments, not only with Sotheby's and the dealers, but also with the fund itself, which didn't always like her suggestions. As she explained, 'My problem with Lewin was that he really did not know about works of art; he felt I had to justify choices in a convincing manner to the British Rail panel, as he would any other investment.'[18] She threatened to resign on three occasions and did not get everything she wanted. When a wonderful Guercino portrait came up at Christie's she was restrained: 'I did a recommendation authorised by Sotheby's Old Master Department, who suggested a moderate bid. I wanted a much higher bid, believing this to be an unmissable opportunity. I passed this to the British Rail panel, who decided to abstain. The picture went to the National Gallery for one bid over what would have been my recommendation. I was furious and sorry at the same time.'[19] On another occasion, Lewin would not let her buy a fragile Degas dancer sculpture. Most works of art she acquired were from Sotheby's, but many came from Christie's and the trade. Edelstein recalled visiting Wildenstein's New York gallery with Peter Wilson, and being astonished by the riches they were shown. While they waited Wilson made a gesture of silence, as he thought that they were almost certainly being bugged.

The fund's purchasing period was roughly five years, and by 1980, when it closed, it had acquired 2,400 paintings and objects at a cost of £40 million (out of a total investment portfolio of over £1 billion). Edelstein had acquired a representative art collection based on three simple principles: to buy the best in every field, to buy from everywhere across the market and not to sell anything for a generation (defined as twenty years). This made it something of a novelty, as most similar investment funds were based on a very narrow base of expertise and were sold too quickly. As far as possible, the fund was established on scientific investment principles.

The purchases were mostly loaned to museums (to reduce overheads), which did nothing to assuage the sense of outrage felt by many, and the fund was certainly the biggest buyer on the London market before the Getty Museum got into its stride. Perhaps their greatest mistake in retrospect – understandable given their conservative policy – was to buy virtually no works dating beyond 1900 on the grounds that they were too speculative. Criticism came from many quarters, mostly complaining about tarnishing creative genius with finance. Geraldine Norman went on the attack in December 1974: 'Was it appropriate for a pension fund to invest in works of art? Was there a conflict of interest for Sotheby's acting as agents for both the vendors and now a buyer? How could their advice be unbiased?'[20] As Lewin himself admitted: 'It was controversial, certainly.'[21]

Among the items the fund acquired were some of the von Hirsch treasures, including the Base for a Candlestick related to the V&A's Gloucester Candlestick. Giuseppe Eskenazi bid on behalf of the fund, paying the enormous sum of £550,000 to secure it, a world record for an English work of art. Further afield, the porcelain dealer Hanns Weinberg bought the Riesener Console Table. Made for Marie-Antoinette at Versailles, it came up for sale in New York and was bought for $400,000.

The fund decided to stop buying in 1981, having spent the £40 million. Its portfolio was reviewed in 1983 and again in 1987, when it was decided to dispose of the entire holdings, albeit much sooner than had been previously envisaged. The pension fund sold prematurely because there was a changing of the guard at British Rail when Christopher Lewin left, and his successor wanted out. There was a feeling that the world had changed, and more appropriate investments were now available. Then there was the impatient desire of the heads of department at Sotheby's to put things in sales, encouraged by a rising market and the appearance of Japanese buyers.

It is sometimes remarked that the British Rail Pension Fund changed the market more through their sales than in their acquisition. The sales began in 1987 and gathered pace. The most successful were the impressionists sold in an evening sale in April 1989, when twenty-five paintings, drawings and sculptures that had been acquired for £3.4 million sold for £35.2 million. The star was Renoir's *La Promenade*, acquired in 1976 for £620,000 and later sold for £9.4 million, eventually ending up in the Getty Museum. This section had grown at a rate of 20.1 per cent per annum, which was 11.9 per cent above inflation. In the same year, the fund's Chinese sale consisted of ninety-six lots, nineteen of which had been purchased from Giuseppe Eskenazi. The item that drew the greatest publicity was a glazed earthenware Tang horse with a pre-sale estimate of £1 million: 'A month before the auction, on a tour of the Far East, the horse was stolen from the shipper's warehouse in Hong Kong, where it had been on show, and was held to ransom.

Eskenazi offered to pay the ransom on behalf of Sotheby's so that the horse would not be harmed. Fortunately, after an intensive undercover operation by the Hong Kong police, the thieves were tracked down and the horse rescued unscathed, in time for the London sale.'[22] The horse made an astonishing £3.74 million, and although the original purchase price was never disclosed, it was probably the highest return that the pension fund achieved. Not all the areas were so successful: tribal art, coins and old master drawings were notably disappointing.

Through its art investments, the fund returned £170 million on an outlay of £40 million. The rate of return, which was over 11 per cent per annum, was 4 per cent per annum more than the rate of price inflation. The fund had outperformed treasury bills but not equities. Despite the controversies that surrounded it, the managers of the British Rail Pension Fund were happy with the result. It had satisfied the main requirement to act as a hedge against inflation. Sotheby's were unquestionably the main gainers, and it can be said that the fund lifted the market at a bad time. It was an extraordinary experiment that has never been repeated on such a scale. Annamaria Edelstein looks back with pride at what they achieved, ruefully commenting that, 'The only people who behaved to me in the same way after it was finished were Eskenazi and Agnew, they continued to invite me to exhibitions and made a fuss of me but then they were the greatest dealers.'[23] Investment in art rears its hoary head every time the market rises, but most sensible collectors would agree that, to paraphrase a well-known saying, look after the art and the investment will look after itself.

Chapter 16

Disrupters: Geraldine Norman and Tom Keating

'As far as I could see, it was the establishment which should be in the dock, not Tom Keating.'[1]

Brian Sewell

The only thing that could unite the two auction house chairmen, Peter Wilson and Peter Chance, was anxiety bordering on paranoia about Geraldine Norman. Iconoclastic, clever, amusing and dogged, Norman created a niche in the art world that makes her the most significant player in this book who was not actually a dealer or auctioneer. The statistician behind the Times-Sotheby's Index, she was subsequently disillusioned with the project. During the assignment, she learnt so much about the art market that she was the natural choice to become the next *Times* saleroom correspondent. Norman was the first saleroom journalist to expand beyond reportage, evincing a determination to bring transparency into the auction houses.

'On 1 April 1969 I started work as Saleroom Correspondent of *The Times*,' wrote Norman.[2] She was to achieve the ambition of Sotheby's publicist, Stanley Clark, of moving regular saleroom reporting from the business pages to home news. But there the satisfaction would end. Norman began to question the statistical basis of how auction houses reported their sales: 'It had struck me as wrong that the salerooms announced a price after every lot, whether they were sold or not – making no distinction between them, and using a fictitious name for those that did not sell … it seemed to me that they [the public] should be told when an item did not sell successfully.'[3] Christopher Wood, ex-Christie's, remembers 'every auctioneer took a list of those [names] with him onto the rostrum. David Bathurst used the Chelsea Football team, but this got him into trouble because he used the name of the goalkeeper, Bonetti, rather too often. The Italian trade became suspicious.'[4] Norman was told by one auctioneer:

> I used the names of beer barons – Trumans, Heinekens, Allsopps, etc. – since beer rarely went with collecting art … I decided that this was a suitable topic for my first feature length article. I went to interview Peter Wilson who told me it was a terrible idea – 'You could destroy the whole art market'. I went to see the chairman of Christie's and met a similar response.
>
> So, I gathered my powder and wrote an explosive draft – not yet printed. The editor, William Rees-Mogg, asked to see my article. He was a keen art collector and generally supportive.[5]

The prospect of the article had the effect of uniting the deadly rivals into action. Wilson and his deputy, Lord Westmorland, joined forces with their Christie's counterparts, Peter Chance and Jo Floyd, to visit Rees-Mogg at *The Times*. As this imposing delegation arrived at the newspaper, Norman recalls how the obituaries editor, 'passing my desk, called them delightedly "four faced villains"'.[6] They set out their belief that the newspaper was damaging the London art trade, making the case that reporting unsold lots as sold was detrimental to the future value of the work of art. When they had finished, Rees-Mogg said quietly, 'I am sorry, gentlemen, but I believe that Geraldine's article serves the public interest. It will be published.'[7]

The article was headlined 'Secrecy in the London Auction Houses', and for Norman, 'It spelled the end of my period of popularity with Wilson, to my distress. I had liked him and admired him a lot. He decided that he could no longer support the Times-Sotheby Index and withdrew his support the following year, using the collapse of the silver market – whose index had disturbingly wiggled down the page – as an excuse.'[8] To Wilson, who had taken Norman up at Sotheby's, this was a betrayal, so, in her words, 'the plug was pulled'.[9] According to Christie's press officer, John Herbert, the failure to persuade Norman and *The Times* to their way of reporting 'resulted in more heated words in Christie's boardroom than on any other subject during my 26 years as a director'.[10]

It wasn't just the reporting of bought-in lots that vexed Sotheby's and Christie's but the investigation into every aspect of their activities. According to Herbert, Christie's board meetings were dominated by the question of what to do about Geraldine, who was reported to have said to somebody from Christie's, 'I hate the guts of both you and Sotheby's'.[11] Norman later denied that she ever hated either firm, but she acknowledged that they had both behaved rudely towards her. The grand head of Christie's Old Master Department, Patrick Lindsay, had apparently rung her up and insulted her for half an hour. Both houses were convinced that she was biased towards the other and were endlessly conducting analysis to prove their point.

The year after Norman's article appeared, Sotheby Parke-Bernet in New York was forced by the courts to reveal its unsold lots and, after an adjustment period, London followed suit.* This would eventually lead to auctioneers saying 'Pass' – a fictitious name was no longer required. Equally important, reserves

* This was in the light of the 'Bathurst affair'. See Chapter 20.

could no longer be set higher than the low estimate. Guarantees on auction items also had to be declared. It was a remarkable outcome.

Norman's next scoop was a more ambiguous matter which ended up in the courts, revealing just how low the art trade had sunk in public estimation. 'The 1970s saw my hour of glory,' Norman recalls, 'with the discovery of Tom Keating', an artist and restorer who had turned into a forger.[12] The saga began with an auction at Arnott and Calver's saleroom at Woodbridge, Suffolk, on 11 February 1970, where a rare Shoreham period drawing by Samuel Palmer (1805–1881) was offered. Palmer was an artist at the height of fashion, as much admired by artists and academics as collectors. The drawing, *Sepham Barn*, was bid to £9,400, a huge price at the time, by Harold Leger, the senior partner of the Leger Gallery of Bond Street. The dealer had rushed to the sale, arriving just in time to find half the art world there chasing the same lot. A few days later, *The Times* received a letter from David Gould, a scholarly dealer and expert on Palmer, questioning its authenticity: 'The reproduction in *The Times* showed sufficient detail to make one wonder if this was a pastiche. The drawing has all the ingredients apart from the somewhat unusual bats ("borrowed" from the 1824 sketchbook?) of a Shoreham period drawing of 1831. But to my eye it lacks the idiosyncrasy which infused Palmer's work with an inimitable poetry.'[13]

Surprisingly, the letter didn't give rise to much response. It wasn't until 1975, by which time David Gould was a cataloguer of paintings at Sotheby's Belgravia, that Norman fell into conversation with him: 'He told me that he had identified several more fake Palmers – did I want to write about them? He introduced me to Raymond Lister who was compiling the catalogue raisonné of Samuel Palmer. And I was able to write about 13

Palmer fakes that were on the market, or had recently been sold as genuine, and query who had made them.'[14] One of these had recently been sold at Sotheby's for a record price.

The Leger Gallery was the centre of the storm. Jane Kelly, Keating's much younger girlfriend, sold four 'Palmers' to Harold Leger for a total of £20,750. Over the next two years, the Leger Gallery sold these on for £40,500. Harold Leger, as ever, relied on his belief in gut instinct and experience over book learning. Keating claimed he had not personally benefited at all from the sales, something that Kelly described as a lie.*[15] Jane Kelly worked on the classic principle of giving information which anybody with knowledge might be able to put together to make the connection back to Palmer. She invented an ingenious provenance which began with the Rev. John Farr, a friend of the artist whose son, Thomas, went to live in what was then Ceylon. His daughter Mary Elizabeth married Douglas Kelly, Jane's grandfather. She told the Leger Gallery that the drawings were sent back from Ceylon in 1967 when her grandfather died. Leger fell for the story. Various people began to question the 'Palmers' but everybody was cautious. There was little clear evidence – just a bad feeling that all was not as it seemed.

Six out of seven of the authorities with whom Norman discussed the Shoreham drawings did not consider them authentic, but they were wary and reluctant to comment. Close analysis showed that the brushwork was not Palmer's, while the paper and pigments were inconsistent and modern. When Norman produced her draft article, the editor William Rees-Mogg took legal advice from James Comyn QC, who gave the chances of winning any court case at about 60/40, with potential damages of about £100,000. Rees-Mogg took the risk and the Keating story kept the nation entertained over the next few years and

* Jane Kelly turned against Keating and did not support his story.

opened a window into the art world, which found the gaze uncomfortable.

On 16 July 1976, Norman published her article about the 'Palmers'. She stated the belief that a forger had been at work and that some of the drawings had come from the same source, Jane Kelly. David Posnett, Harold Leger's young junior partner, protested vigorously in *The Times* letters column to defend his firm's integrity.[16] This provoked a considerable correspondence both for and against. The trade bodies, BADA and SLAD, convened inquiries on the subject but were divided in their response, uncertain if it was the honour of their members, or the public, they should be protecting.[17] After Norman's article appeared, she started receiving telephone calls informing her about a Suffolk artist, Tom Keating. She followed the trail to Dedham, where she found his cottage and knocked on the door. It was opened by a man in his sixties with a white beard and large expressive eyes and, as Norman recalled, 'I thought I had met Father Christmas.'

Keating had an engaging, twinkly manner, and Norman described him as, 'A jolly cockney, house painter turned artist. He had been allowed to attend Goldsmiths College at the state's expense after spending the war in the royal navy. Among the superior, cultured students he was always conscious of a lack of "taste". He began copying great artists in the hope of acquiring "taste". Keating became a picture restorer and on the side had flooded the market with pastiches of Degas, Constable and the Canadian artist, Krieghoff since the early 1960s.'[18] He felt he had not been given his due by the art establishment and consequently had a grudge against dealers: 'I flooded the market with the "work" of Palmer and many others not for gain ... but simply as a protest against merchants who make capital out of those I am proud to call my brother artists, both living and dead.'[19] His stated aim was always to make the art world look foolish, and in this he succeeded. Keating claimed to have produced between 1,000 and 2,000 'Sexton Blakes', as he described them in cockney rhyming slang, including eighty 'Palmers'.[20]

Geraldine Norman enjoyed disrupters, including Peter Wilson and the mischievous David Carritt, and she had a soft spot for ambivalent antiquities dealers like Bob Hecht. Her cockney husband, Frank Norman, had spent time in prison and wrote the musical *Fings Ain't Wot They Used t' be*, so she was predisposed to sympathise with Keating. He was everything she could hope for: a cockney artist of great warmth with a grudge against 'the smooth men of the art world', the underdog who barked. As far as the art world was concerned, Norman was, in the words of one eminent dealer, 'a highly intelligent, fascinating, left-wing character and stirrer'.[21] Part of Norman's problem was knowing how much she could accept of Keating's account. He enjoyed a good yarn, and his daughter confessed that 'Dad's always been a bit of a romancer'.[22]

According to Keating, the Redfern Gallery in Cork Street had bought sixty modern drawings in a job lot from a Kew junk dealer, mostly in the style of German expressionism and French impressionism, and subsequently sold some of them as genuine. The Redfern records showed that, in 1963, they had purchased approximately thirty works which the gallery regarded as a mixed bunch of geese and swans. One, a 'Mondrian', was almost sold to the National Gallery of Scotland but went to a Belgian private collector instead. When it was returned after doubts were expressed, Redfern began to have reservations about the whole cache and wrote to all the buyers offering to refund them. It was the Redfern Gallery, followed by the Leger Gallery, who made the first complaint to the police which set the court case in motion.[23]

Norman's second article, which revealed Keating's name, came out while the artist was touring the West Country on his motorbike. He picked up a copy of *The Times* at Glastonbury and called Norman asking, 'What do I do next?' 'I told him that he'd got to tell what he'd faked and why, explain his life. I said that I'd help.'[24] She instructed him to catch a train and come and have dinner with her that night. This took place with her husband, Frank Norman, at a trendy restaurant, Odin's. By the end of dinner, it was decided that Frank would help Keating write his life story, and Geraldine would check it and supply an introduction. They took Keating with them on holiday in Berkshire the following week and the book came out in 1977, entitled *The Fake's Progress*, a reference to Hogarth's famous series of paintings. The story made Norman 'News Reporter of the Year'.

'The reaction of the art dealing community,' in Norman's view, was 'basically antagonistic'.[25] They were at the very least defensive, and the story certainly dented what she described as the aura of invincibility and respectability of the upper echelons of the trade. The auction houses' reaction echoed the dealers, and constituted 'a thorough dislike of the matter being publicised, a reluctance to assist further than answering very specific questions, and groans at the suggestion that they might have handled Tom's work' (they all did under various cataloguing descriptions).[26] Leger questioned the scientific evidence concerning 'Shoreham Moonlight' when the owner, a Mr Green, asked the gallery to take the picture back. When Leger refused, he went to BADA, who set up an arbitration machinery, after which Leger refunded him the money.[27]

Tom Keating and Jane Kelly were both arrested at the end of July 1977, a month after the publication of the book. The charge against Keating was that he had conspired with Kelly to mislead the Leger Gallery into believing that works by him had been painted by Palmer. Kelly's creation of an ingenious and fake provenance was central to this deception. The court case opened in January 1979. The defence counsel, Jeremy Hutchinson QC, spent his brilliant career at the Bar defending those whom he perceived as underdogs. He liked Keating and described him as being 'as close to an anarchist as any man I ever met'.[28] He also accepted the artist's plea that his motivation was 'getting my own back on dealers who exploit people in this way'.[29]

Hutchinson's approach was to 'arouse in the minds of the jury a suspicion that the greedy dealers were well aware that the works might not be genuine, but that the possibility of making a substantial profit overcame their scruples, operating as they were in a dubious market'.[30] *The Observer* described the Old Bailey trial as 'the best show in town'. David Posnett, the young Leger Gallery director, took much of the brunt of Hutchinson's devastating sallies and complained that Keating had made Leger 'a laughingstock of the trade'. One surprise witness for Keating was Brian Sewell, who described his intention 'to pour cold water on the art establishment. As far as I could see, it was the establishment which should be in the dock, not Tom Keating.'[31] That was in effect what happened. In some ways, the trial resembled the Peter Wright *Spycatcher* trial in Australia (1987), in which the ins and outs of the case were subsumed under the guise of poking fun at the British establishment.*

The case against Keating never reached a verdict. Out on his motorbike, Keating skidded on ice and was badly injured. The judge adjourned the trial, discharged the jury and dropped the prosecution – it was never clear why, though the case had already dragged on for weeks and had taken a toll on Keating's health. Hutchinson felt a certain sympathy for the Leger Gallery against whom he had scored so many laughs, and he acknowledged that the pastiches were very fine (a judgement few would agree with today). The artist presented Hutchinson

* The trial held in Australia was an attempt by the British government to prevent Peter Wright's book *Spycatcher* (1987) being published. It was a lamentable failure, but entertained the world mightily.

with a Constantin Guys pastiche, which he had sketched during the trial, that he hung in his office. Jane Kelly was sentenced to a short prison sentence.

For Geraldine Norman, the Keating affair was a cracking scoop. She knew she had a great story: the disgruntled unrecognised artist, as she saw it, taking his revenge on the grandees of the West End art world whom he had conned. Her conclusion was, 'That the general public has unquestionably enjoyed the Keating affair. There is something which appeals to the anarchic strain in all of us about the "experts" – those pompous connoisseurs who are always putting us in our place – being taken for a ride.'[32] To the art trade, this was another example of her malevolence towards them. The main lesson taken from the scandal was that gut instinct such as Harold Leger's was no longer sufficient. It meant a more stringent approach to matters of authenticity and cataloguing. A second lesson was that being defensive under scrutiny was not the best policy. The main plank of Hutchinson's attack on the trade – focusing on greed and deliberately gulling the public – was not entirely fair. The Leger Gallery had given Jane Kelly a good price, and a 100 per cent mark-up would have been normal for a West End gallery. It was, above all, a case of poor connoisseurship.

After the scandal, Keating prospered: in 1982, Channel 4 gave him a series explaining how to paint in the manner of the old masters. Such was the hyperbole surrounding the artist that Magnus Magnusson thought that 'Tom Keating was almost on a par with art historian Kenneth Clark and his pioneering 1969 BBC television series *Civilisation*.'[33] Art forgers tend to be treated as folk heroes, and Keating was portrayed as a lovable rogue, a depiction the art trade could never accept. Despite that, the year after his television debut, he was given his own sale at Christie's South Kensington. Keating was fortunate to have had a brilliant publicist and apologist in Geraldine Norman and a great barrister, Jeremy Hutchinson, to defend him. To the credit of David Posnett, not only did Leger Gallery survive the scandal but, after Harold Leger's retirement, he took the firm into the first rank of British painting dealers.

Chapter 17

The Chinese Market

'Chinese art will see me out, but there is no future in it.'[1]

Peter Sparks to Anthony du Boulay

London held the central position in the Chinese art market until the late 1980s, when Hong Kong, and later mainland China, reduced it to secondary status. In the 1920s and 1930s, London had become the leading centre for the Chinese art trade outside China. The landmark formation of the Oriental Ceramic Society (OCS) in England in 1921 brought about a remarkable nexus of museum curators and wealthy collectors. The most famous of the collectors was Sir Percival David, whose collection of flawless Imperial wares is on permanent loan to the British Museum. The OCS was one of the most beneficial conjunctions of scholarship and trade in the history of British collecting and contributed to the great enrichment of national collections. All the necessary components developed in tandem: brilliant collectors, strong academic and museum cooperation, and, above all, a roster of sophisticated dealers including Bluett, Sparks and Spink.

During the period covered by this book, the auction houses were the market makers, especially Sotheby's, which created the Hong Kong sales that paved the way for mainland China's participation. No market was more international than the Chinese, and Roger Keverne pinpointed the waves of influence: 'Japan in the late 1960s, Hong Kong in the 1970s, Taiwan in the 1980s and mainland China in the 1990s.'[2] Two names dominate the London trade: Julian Thompson (1941–2011), the head of Sotheby's Chinese department (and sometime chairman of the firm), and Giuseppe Eskenazi, described as 'The leading dealer at the highest level in Chinese works of art in the world'.[3] They were very close and together virtually ran the market.

Thompson's mentor was Jim Kiddell, who had joined Sotheby's from the Ministry of Pensions in 1921, knowing nothing about art. During his fifty-eight years with the firm, he made himself the specialist in Asian art – amongst other things – and set a high standard of cataloguing. Julian Thompson arrived for a summer job in 1962, while reading maths at Cambridge. Unsure what to do with him, Kiddell gave him a group of nondescript items to catalogue, which he did so beautifully that it was clear that he was going to be a star. Five years later, Thompson became head of the Sotheby's Chinese department, and his influence on the international market was soon to be felt. Shy and cerebral, Thompson was universally acknowledged as the leading expert in the trade on Chinese porcelain. To Giuseppe Eskenazi, 'He was by far the cleverest person I worked with, for his intuition and sharp eye.'[4]

Anthony du Boulay joined Christie's in 1949 when it was still at Spencer House after the wartime bombing of its King Street premises. Aged nineteen, du Boulay was put into the decorative arts department, which dealt with furniture, porcelain, glass and antiquities. He ignored the advice of the distinguished dealer Peter Sparks, that the Chinese market had no future, and in 1956 formed a separate Ceramics Department when he sold the Fuller sale of Ming 'Blue and White' porcelain. Du Boulay explained that Christie's gained most of their material in those days from grand country houses, while Sotheby's held more collector sales, typically for Lord Cunliffe in 1952 and Robert C. Bruce the year after. During the 1960s, du Boulay recalled, export porcelain and *famille rose* were still very fashionable and essential components of furnishing grand British houses.

The availability of jet travel and the rise of prosperity in Japan in the 1960s brought a contingent of well-funded Japanese dealers to London auctions, ready to pay higher prices for the best Chinese art. This competition brought pieces from old collections onto the market, creating something of a boom by the end of the decade. The beginning of the rise in prices of Chinese art is sometimes dated as early as 1963, with the Sotheby's sale of the Museum Street dealer H. R. N. Norton. This offered a broad range of items from Tang to Qing ceramics, jade, bronzes and lacquer. It was Christie's who mounted one of the most celebrated auction sales at the height of the boom, Chinese art from the wide-ranging Alfred Morrison collection, removed from Fonthill (1971). Morrison had bought his most significant pieces in an unusual way. In 1860, Captain Henry Brougham Loch, posted to Peking as part of Lord Elgin's embassy, was imprisoned. When he was released, Loch acquired items sacked from the Summer Palace, known as the 'Garden of Perfect Brightness', in Peking. He sold them to Morrison in a London club (said to be the Travellers') after establishing that he was the wealthiest man in the room.[5]

Japanese buyers concentrated on ancient Chinese pottery, archaic bronzes, Yuan and early Ming porcelains and lacquer wares. Both Sotheby's and Christie's attempted to mount public auction sales of Chinese art in Tokyo during the late 1960s and early 1970s, but the dealers there united in opposition to the introduction of auctions run by foreigners.* Raising a boycott, they quickly blocked the plan. The Japanese market was an equally closed shop between fellow dealers, but it was possible with patience and good manners to acquire material from some

* Christie's succeeded in holding Asian art sales at the Hotel Okura in Tokyo in February 1980 and February 1981.

in the Tokyo trade. When Giuseppe Eskenazi wanted to do business there, he took advice from the British ambassador, Sir John Pilcher, who explained the complex and formal etiquette involved in visiting Japanese collectors. Objects were shown one at a time, partly to enable the visitor to appreciate their beauty, and partly in order that the owner might judge whether the visitor was worthy of being shown another item. Japan was a treacherous market for London dealers, and Eskenazi nearly went bankrupt owing to the sudden absence of Japanese buyers in the late 1980s. The Japanese virtually withdrew from the international art market again when the financial bubble burst in 1991.

Although extraordinary collectors like Eiitchi Ataka in Osaka would only buy and sell through local Japanese dealers, one of the few Japanese collectors who came in person to auctions at Sotheby's and Christie's was the colourful Seijiro Matsuoka, who started buying in the 1970s. He attended his first public sale at the age of seventy-eight and was a bit of a show-off. Speaking little English, in the days before currency converter screens it was a mystery how he knew what he was paying. When asked, he simply replied, 'My hand up long time, high price. Short time up, cheap price.'[6]

The year 1973 was one of change: the oil crisis, the first Sotheby's sale in Hong Kong and the collapse in the Asian ceramics market, exacerbated by the Portuguese Revolution, set the scene for change. Until then, prices for Chinese ceramics had been rising steeply at London auctions, stimulated not only by the Japanese but also by two aggressive dealers: Hugh Moss and Helen Glatz. Of German origin, Glatz had come to London after her release from a concentration camp, and was described as taking 'the London market by storm as a dealer between 1968 and 1974 on behalf of a wealthy Portuguese client, Jorge de Brito'.[7] When bidding, Glatz would simply hold her biro vertically in the air. Consequently, she ran up massive debts at Sotheby's on de Brito's behalf, buying Chinese art, particularly Ming ceramics.[8]

Hugh Moss, equally invested in the Portuguese market, was something of a pioneer who revitalised his family firm. His father, Sydney Moss, had a very high reputation, although if something was described as 'a bit Mossy', to the trade it meant that it was a bit dull. Hugh was a more dynamic figure, widely regarded as the best judge of Qing dynasty enamelled porcelains, especially those inscribed with imperial reign marks. He also played a key role in establishing the collecting category 'Chinese Art from the Scholar's Studio' – including brushpots, ink stones, brush washers, seals, ink cakes, scholar's rocks and other paraphernalia of the scholar-elite in China. Moss set up a City investment group with the financier Jim Slater. Buying Chinese art in the rising market, the group became a casualty of the downturn.

The economic impact of the oil crisis and the Portuguese Revolution in April 1974 left Sotheby's with £7 million of unpaid sales, and Christie's with £1 million. Moss's business crashed and he moved to Hong Kong to rebuild: 'I left London in 1975, mainly for economic reasons, after the oil crisis and subsequent crash of the art market. My move to Hong Kong came at a time when the centre of the Chinese art market was shifting eastward, and Hong Kong rapidly became a fourth major centre for Chinese art along with London, Paris, and New York.'[9] His move to Hong Kong was astute, and he revived his business as the market reawakened.

The establishment of Sotheby's Hong Kong auctions in 1973, followed by Christie's the next year, became a roaring success. Traditionally, the problem with the Hong Kong market had always been the lack of expertise and the number of fakes being offered on Hollywood Road, the main dealers' thoroughfare. With the expertise of Julian Thompson and his deputy, Jim Lally, the arrival of Sotheby's in the city was transformative and allowed buyers to trust what they bought. As Giuseppe Eskenazi explained, 'Auctions solved the problem by making genuine material available to a much wider audience and by providing market information through publication of pre-sale estimates and price lists.'[10] Fakes were the one thing that made Julian Thompson, normally calm and placid, angry, and he would want to smash them. Such was his reputation that notices appeared in Hollywood Road shop windows: 'Julian Thompson has authenticated this'.

The Hong Kong auction market developed slowly at first, but by the 1980s a group of talented local collectors emerged, notably T. Y. Chao, C. P. Lin and Au Bak Ling, advised largely by Thompson.* The doyen of Hong Kong dealers, Edward (E. T.) Chow, is an almost legendary figure today.[11] On his death, Chow entrusted Thompson and Michel Beurdeley to sell his personal collection. The 1980–1 sales that followed were, by any standards, a landmark. Parts I and III, Ming and Qing, were held in Hong Kong, while part II, pre-Ming and bronzes, went to London. These sales nourished the next generation of Hong Kong collectors, and their success signalled the return of confidence into the market. With the huge exception of Giuseppe Eskenazi, London would now play a supporting role. The period from the end of the 1980s saw the demise of many great old London firms.

Spink & Son, established in 1666, was usually described as the oldest antique dealer in the world. The firm was a Royal Warrant Holder and had a large general dealership that sold Asian works of art, silver, coins and medals (in which they were

* Au Bak Ling, a brilliant collector but the most difficult of all customers, had an exhibition of his collection at the Royal Academy, *100 Masterpieces of Porcelain*, in November 1998. A catalogue was written but Mr Au refused to allow it to be published, probably because he did not want attention from the Chinese tax authorities. He was nocturnal, and the author was once shown his collection at 4 a.m.

paramount), alongside furniture and paintings. Spink were the closest thing to Harrods in the art market, with the widest scope of any dealer in Europe. Chinese art formed about 30 per cent of their business. Their excellent Asiatic art department had a broad range of objects, including Qing, cloisonné enamels and jade, and they were ahead in the taste for bronzes. Unlike Moss, however, Spink failed to adapt, and, as Eskenazi described, 'the firm remained static and had far too much material on approval. They did not understand the new Japanese and Chinese buyers; they had better expertise on India.'[12] Spink's days were numbered when Christie's bought it in 1993, largely for the real estate, and that kind of business had probably had its day.

The main legacy of Spink's Asian department was Roger Keverne. Roger (like John Hewett) came from a family of Plymouth Brethren. Starting his career at Spencer's of Retford, he was appointed director of Spink's Asian department at the age of twenty-eight, a position he held for seventeen years, serving royal patrons as well as collectors like Sir Michael Butler and the Rockefeller family. In 1992, Keverne started his own business with his sinologist wife, Miranda Clark, specialising in Chinese antiques, lacquers, jades and cloisonné. One of their proudest achievements was to help assemble the extraordinary collection of archaic bronzes at Compton Verney Museum, established by the British philanthropist Sir Peter Moores. Until he retired in 2021, Keverne tirelessly served on boards and committees, as president of BADA, as well as twice being chairman of 'Asian Art in London'.

The Marchant family, based in Kensington Church Street since 1950, have always kept one step ahead. As Roy Davids and Dominic Jellinek observed, 'Through the 1950s the London dealers were receiving shipments from Hong Kong, and beginning to make visits there themselves, with Sydney Marchant believed to be in the lead.'[13] Sydney's son Richard joined the firm in 1950 and gained a reputation for aggressive bidding.

As Gerald Reitlinger said to him one day, 'Marchant! You know what your trouble is? You bid like a collector!'[14] Richard Marchant was one of the first dealers in London to concentrate on Qing, which turned out to be where the Chinese market was heading, and consequently the company thrived and is still in business.

Bluett & Sons, founded in 1884, was a scholarly firm which held exhibitions and published catalogues. They kept the OCS tradition of Percival David alive, fostering the taste for Imperial ware, and had the good fortune to be able to recycle collections they had helped to form in the 1920s. From 1964, the firm was run jointly by Roger Bluett and Brian Morgan, who embodied a slightly old-fashioned academic taste for very high-quality pieces. The atmosphere of the shop was like a gentlemen's club. Influential, knowledgeable and loved by clients, Morgan was mentor to Dominic Jellinek, co-author of *Provenance*, the standard work on anglophone Chinese art collectors and dealers.

Bluett's exhibitions were eagerly anticipated. At the opening of a 1971 exhibition of ceramics from the Cunliffe collection, a notorious incident occurred. The collector Frederick Knight, at the best of times capable of erratic behaviour, wanted a Chenghua 'Palace' bowl and joined the queue outside the gallery on the morning of the opening. In front of him was a student who stood there until a Filipino gentleman appeared from Claridge's Hotel. The new arrival handed the student a package of money, and brazenly stood in front of Knight, who sensed that they were after the same piece. When the door opened there was a scuffle, in which Knight elbowed his competitor out of the way and came away with the bowl. Roger Bluett intervened at this point and the Filipino gentleman warned Knight threateningly never to visit the Philippines.[15]

Bluett acted for several Japanese dealers at auction, but it did not bring them much profit, and they received merely a small commission on the sales. They had no access to private collectors in Japan. As one observer told the author, 'The truth is that Bluett was less interested in money and often sold at far too small a profit.'[16] After 108 years, the firm finally went out of business in 1992.

Another grand old firm with a similar lifespan to Bluett was John Sparks Ltd (1888–1990). Percival David was a client of both firms, and Sparks also had several royal clients, notably King Gustaf VI Adolf of Sweden. The shop was in Mount Street, where the accent was showier and more decorative than Bluett, with a good stock of *famille rose*. Peter Sparks, who was friendly with Queen Mary, was by all accounts very suave. One day he brought the Queen back to the shop after lunch at the Connaught Hotel. His partner, Peter Vaughan, had been drinking heavily and collapsed in front of Queen Mary, who behaved as if it was perfectly normal and probably the result of the hot weather. When he took over the firm, Vaughan suffered from that disease of believing you are as rich as your clients. He drove a convertible Rolls-Royce, usually with his two pet Chow Chows – whose colour matched the car – and would lunch every day at the Connaught. Perhaps the firm's most visible legacy is the Duberly Collection, a Sparks-formed assemblage bequeathed to Winchester College in 1978, and a perfect microcosm of taste from the 1950s to the 1970s.

By far the largest auction sale of Asian art in modern times was the Nanking Cargo sale at Christie's in 1986, sold in Amsterdam but organised from London. The cargo from the Dutch ship *Geldermalsen*, sunk in 1752, was found in 1985 by British-born salvage expert Michael Hatcher. It made a sale of 2,800 lots, and attracted over 125,000 absentee bids. The sale totalled over £10 million, helped along by the policy of dividing the lots into usable dinner services for twelve. One lot demonstrates the mania that surrounded the sale, as the organiser Colin Sheaf recalls, 'the estimate was £3,500 but the item made £8,000. The dealer who purchased it resold it almost immediately for £15,000 and only a few weeks later was offering his client £30,000 pounds in order to resell it for £60,000.'[17]

The Italian Giuseppe Eskenazi was widely recognised as the leading Chinese art dealer in London and, some might say, the world. His annual exhibitions, lavishly catalogued, drew people from around the globe. When Sir Isaiah Berlin was asked whether he would like to visit China he replied, 'Only if Eskenazi came along!'[18] Born in Constantinople in 1939, Eskenazi learned to speak French, Greek and Turkish before he learned Italian. The family art-dealing business was established in Milan in 1923 by Giuseppe's uncle. Educated in England, Eskenazi set up in London initially with a view to supplying works from the British market to the Milan office. He found a mentor in Sydney Moss, who explained the dating of porcelain and how to distinguish fifteenth- and eighteenth-century versions: 'He taught me how to look and what to look for.'[19] Otherwise, 'the English dealers didn't want to know me or care for me', and as a result he found most of them rather pompous: 'Bluett and Spink looked down on me.' Eskenazi was encouraged, however, by Sotheby's chairman, Peter Wilson, who said he would help him open a gallery. True to his word, Wilson put the Sotheby's PR machine at his disposal for his first exhibition, and the result was that Geraldine Norman wrote a full-page article about it in *The Times*. From then on, Eskenazi's business model was to put together an exhibition every year.

Early in his career, Eskenazi wanted to exhibit at Grosvenor House, but two or three top dealers kept him out. To make matters worse, they would visit his small office the week before the sale and purchase his best items, which he would then find on their stands, all marked with a red spot. This both eliminated him as competition and provided them with good material. As a result, Eskenazi avoided fairs and preferred to do his own exhibitions – he is probably the only major dealer who never participated in a fair. His hero and greatest friend in the trade was Julian Thompson: 'We never had a quarrel or disagreement. The fact is he was never wrong. He had a brilliant eye and absolute honesty.' As a result, Eskenazi was '100 per cent Sotheby's in his allegiance to Thompson'.[20]

As he told an interviewer in 1972, 'my great ambition was not to have a shop … I know what a waste of time it can be. I wanted a gallery where you see people by appointment – curators, dealers, collectors to whom you don't have to explain what a Tang pot is. I spent 99 per cent of my time persuading people to sell and 1 per cent persuading them to buy. Good things sell themselves. I have one great collector and client: myself. My collection is here.'[21] Eskenazi had a novel way of dealing with who should be given first option on items from an exhibition. He introduced a scheme whereby the first week was set aside for examination by potential purchasers. Any offers above the published base price were submitted to a sealed container, and on a prearranged day the seal was opened by an independent lawyer, who announced which slip had the winning bid.[22]

When the market collapsed in 1973–4, Eskenazi realised that the only thing to do was start buying again at the new, much lower level, negotiate a bank loan to make this possible and put away the expensive unsold stock. In 1975, he held an exhibition of bronzes from the celebrated Stoclet collection in Brussels. Bronzes were not popular at the time – as the Taiwan collector Bob Tsao told the author, 'they were large, heavy and often dirty'. Eskenazi was the first dealer to make a full catalogue of bronzes. He was also at the forefront of the new fashion for pottery tomb sculpture. The taste for such archaeological material developed slowly as, initially, neither the Chinese nor the Americans wanted to buy it.

In his memoir, Eskenazi recalls several encounters with famous collectors. John D. Rockefeller one day asked 'what discount do I get?' Giuseppe laughed, which disconcerted Rockefeller, and Giuseppe explained that, 'in Europe the name

Rockefeller is synonymous with wealth, so asking for a discount sounds a little out of place'. Rockefeller responded, 'Yes, but everything I save on this purchase will help me on the next purchase from you!' The discount was given.[23] Another collector who appeared in Eskenazi's gallery was Ronald Lauder, who explained that he divided art into three categories: the 'O', the 'Oh My' and the 'Oh My God', making clear that he only wished to see vessels belonging to the last category, which he duly did.

Collectors sometimes appear from nowhere. In 1997, Peter Kranz, a Swiss gentleman living in New York, showed up in the gallery with a copy of Sotheby's *Preview* magazine.[24] He pointed to the illustration of a beautiful Qianlong bowl decorated with landscape scenes coming up in Hong Kong. Kranz knew nothing about Chinese porcelain but asked if Giuseppe thought it was good and whether he could buy the bowl on his behalf. They paid HK$21,470,000 (£2,158,712) for it and Kranz was a rare case of a collector starting at the top, or, as Giuseppe put it, 'starting with the crown and adding the orb, sceptre, and other jewels ever since'.[25]

In the spring of 1993, Eskenazi moved his business to an Edwardian mansion in Clifford Street. It was about as inauspicious a time as you could imagine, following Black Wednesday of September 1992. Many leading and famous old names in Asian art in London closed over the next year. Recognising where buyers were coming from, Eskenazi was the first London dealer to do catalogues in both Chinese and English. In 1999, he was able to acquire one of the crown jewels of Chinese porcelain, a *doucai* chicken cup for HK$26.5 million (£2,664,815), formerly in the collection of Mrs Leopold Dreyfus – 'Mrs Chicken Cups', as E. T. Chow dubbed her – who owned two of them. Today, the firm continues under Giuseppe's son Daniel, who started working for his father in 1990 after a stint at Sotheby's in New York.

In Giuseppe Eskenazi's opinion, the demise of dealers like Sparks and Bluett was 'due to their preference for purchasing objects on commission and selling goods on consignment rather than buying for their own stock. It is very difficult to run a successful gallery on commission only.'[26] By the 1990s, the European market in Chinese art was virtually owned by Eskenazi in London and Gisèle Croës in Brussels. However, the rising market in Hong Kong, Taiwan and mainland China had eclipsed the rest of the world. Since the beginning of the millennium, New York and London have become relatively minor participants in the international Chinese art market, lacking new collectors willing to compete at the higher price levels.

While Sotheby's and Christie's obsessed over mainland China, by 2000 probably the best Asian department at a London auction house was at Bonhams, which had become a major player in the Chinese market. Alongside this, a regional auction house, Woolley and Wallis of Salisbury, had good Chinese expertise and was becoming a force, attracting sales away from the London salerooms. Eskenazi lamented the absence of serious British buyers of Chinese art since the 1990s, compared to the giants between the wars. However, to him the important point is that as long as expertise remained in London, the city would survive as a leading centre, albeit one dependent on overseas visitors.

Chapter 18

Victoriana

'One of the most amusing changes in sale-room fashion is the advent of the Victorians.'[1]

Lillian Browse

When Sotheby's opened their saleroom dedicated to Victorian art in Belgravia in 1971, the Grosvenor House Fair still refused to admit anything made or painted after 1830, or even accept that it was 'antique' or indeed 'a work of art'.[2] Jeremy Cooper described the imperceptible pushing forward of the date lines, from Georgian, to Regency, to William IV – anything but Victorian, a word which, as Cooper recalls, 'was at this time pronounced with the same note of disdain as Lady Bracknell exclaiming "A handbag!?"'[3] Victorian art may have been a pejorative term – Bloomsbury intellectuals were particularly contemptuous of it – but, as Rupert Maas reflected, 'The period covers three generations, from Britain as a rural world to an urban world, and there is so much variety and it is so multifarious that its art never entirely fell into disrepute. Certain artists early in the reign, like Turner, and later artists, like Burne-Jones, were always admired. It was the big declamatory paintings by Alma-Tadema that went out of fashion and anything that was shouty.'[4]

The gradual revival of Victorian art and architecture took many forms: the writings of John Betjeman, the cartoons of

Osbert Lancaster, Peter Floud's seminal 1952 V&A exhibition *Victorian & Edwardian Decorative Arts*, the founding of the Victorian Society in 1958 (and its successful later battle to save St Pancras Station), more frequent exhibitions of Victorian painters, decorative arts and sculpture, the interest of the colour sections of the Sunday magazines and films set in the period, like *Oliver!* (1968). Anybody interested in architecture joined the Victorian Society, where they would meet like-minded young scholars such as Mark Girouard and Clive Wainwright. The 1969 Housing Act gave grants for the first time to restore old houses, which resulted in a vogue for restoring Victorian terraced housing. New buyers wanted suitable antique furniture, and fabrics by Laura Ashley, founded in 1953.

The central strand in the rise of Victoriana, however, was the revival of interest in the Pre-Raphaelite Brotherhood. Christopher Wood, the future head of Christie's Victorian Paintings Department, remembered trying to view pre-Raphaelite paintings at the Tate during the 1960s, 'a group of their pictures were then to be found huddled together in a small room in the basement next to the gents'.[5] Wood observed that, at this time, expertise in 'the pre-Raphaelite field was dominated by a quintet of formidable ladies – Diana Holman Hunt, Rosalie Mander, Mary Lutyens, Mary Bennett and Virginia Surtees'.[6] A landmark was the publication of Graham Reynolds's book, *Victorian Painting* (1966).

Aspects of Victorian painting had remained popular within the art trade. Gerald Reitlinger noted that the middle rank of nineteenth-century coaching and hunting pictures by Ansdell, Ferneley, Herring and Pollard 'were dearer in the 1960s than those of such eminent Victorians as Landseer, Leighton, and Alma Tadema. They could be dearer too than anything by Rossetti or Burne-Jones.'[7] There was also the evergreen market for salon and genre pictures beloved by North London collectors and supplied by firms like Coolings, Newman & Sons and MacConnal-Mason.

Julian Hartnoll, one of the most important dealers in the pre-Raphaelite revival, describes the type of picture:

> It was an unacademic selection from the nineteenth century, English landscape scenes by B. W. Leader, F. W. Watts, the Shayers, a group of the Norwich School painters, and Birket Foster of course; from France some of the Barbizon painters; from the Netherlands the ever-present Koekoeks and Leickert; flower paintings by Ladell and Fantin-Latour. Figure subjects were restricted to shiny period costume pieces of laughing cavaliers by Roybet, carousing cardinals by Croeghart or Brunery, and eighteenth-century pastiches by Marcus Stone and Frith, Italian girls by Eugene de Blaas, Cromwellian soldiers by the nearly contemporary F. M. Bennett. Where though were the serious painters? No Leightons, no Poynters, no pre-Raphaelites. It was almost as if anything which was too obviously Victorian was to be avoided. Nothing was to disturb the owner or challenge the customer intellectually.[8]

The partners of Newman & Sons told Hartnoll, 'Frankly, we don't understand the pre-Raphaelites.'[9]

In Hartnoll's opinion, the most serious interest in Victorian art after the war was shown by the Leicester Galleries, which put on seminal shows during the 1940s, culminating in *The Victorian Scene* in 1956. Another early pioneer from the 1940s was Charlotte Frank, who handled some of the finest and best examples, operating from a basement beneath Berry Bros in St James's Street. She had a group of discriminating clients, notably the diplomat Sir David Scott, who first entered her premises in 1947 and would form the finest collection of narrative paintings, which was more a survival of taste than a revival.* Charlotte and her husband Robert had come to London just before World War II as refugees from Nazi persecution.† Charlotte handled great paintings like John Martin's *The Plains of Heaven* and *The Last Judgement*. That singular peer, Lord Lambton, would saunter down from White's after lunch to buy Waterhouse paintings from her. Frank also sold works by European symbolists and became the preeminent dealer in the pre-Raphaelite followers that John Christian later labelled 'The last Romantics'.[10]

Continuity was provided by the Fine Art Society, and in a peculiar way by Agnew's, who had been the original dealers for many of the top Victorian painters they still sold. Founded in 1876, the survival of the Fine Art Society was rather a miracle. For them the taste for Japan and the aesthetic movement had never gone away, and its contribution to the Victorian revival was as much about the decorative arts as painting. A series of paradigm-shifting exhibitions and catalogues highlighted a taste inclined to the Arts and Crafts Movement and Charles Rennie Macintosh cabinets rather than pre-Raphaelite paintings. The 1969 exhibition *The Earthly Paradise* included major works by then comparatively unknown artists like Cayley Robinson and Joseph Southall. Among those who bought from the exhibition were Charles and Lavinia Handley-Read, who lived in a house in Ladbroke Road crammed with sculpture, paintings and nineteenth- and early twentieth-century furniture. Charles Handley-Read's dictum was 'away with the aesthetic gap'. Writing about him, Christopher Wood recalled, 'At last I had found a true Victorian collector, not only older than me, but incredibly serious about it … Charles's God was William Burges and he already owned several pieces by him, including

* Scott remembered going to the 1895 RA exhibition aged eight and admiring a Leighton. The sale of his collection was held at Sotheby's in 2008.

† Anne Frank was Robert's niece.

his bed from Tower House in Melbury Road.'[11] When he saw these pieces through the window of the Lacquer Chest, a shop in Kensington Church Street, he queued all night on the pavement to be in first when it opened the following morning. The Handley-Read collection was dispersed by the Fine Art Society and Sotheby's, but there is a Burges room at the Cecil Higgins Art Gallery in Bedford dedicated to them both.

The Fine Art Society served another first-rank collector, Christopher 'Kip' Forbes, who formed a glorious collection of works by Millais, Burne-Jones and Whistler that hung in Old Battersea House, an eighteenth-century London manor house associated with the pre-Raphaelite painter Evelyn De Morgan and her husband, the potter designer William De Morgan. Forbes enjoyed teasing his father, Malcolm, who owned a Monet *Water Lilies*, that his entire collection of Victorian art cost about the same as this one painting.

The dealer who perhaps more than any other brought Victorian art into the mainstream was Jeremy Maas (1928–1997). Wood described him: 'Very tall, reserved, pipe-smoking, and urbane, Jeremy hardly seemed to fit in his very small gallery … he was always full of new stories about Victorian artists and I well remember him telling me between excited puffs on the pipe that "Mulready never drew a female model unless he had slept with her first."'[12] Maas always attributed his discovery of Victorian painters to the fact that he had not studied art history. Instead, he had read English at Oxford, where there was no art history course, reflecting that, 'Had there been a proper course I would have been steered away. The interest would have been driven out of me. But I remained undefiled.'[13]

After a spell in printing and then advertising, Maas set up Bonhams drawings department. He opened his own gallery in 1960, and the following year held his first show of pre-Raphaelite paintings. Two years later, he pulled off what is now regarded as his greatest coup, buying Lord Leighton's then almost unsaleable

Flaming June for £1,000 from a man who had bought it from his hairdresser for £60. Maas managed to sell it to Luis Ferré, who acquired it for his museum at Ponce in Puerto Rico for £2,000.* Today it is probably the most admired Victorian painting outside Britain. A literary agent's suggestion that the painting would make a wonderful book cover was the genesis of Maas's *Victorian Painters* (1969). Among the book's pioneering aspects was Maas's recognition of fairy paintings (d'Offay had an exhibition of them in the 1960s). However, the gallery could only survive by selling paintings (including Constable and Turner) to Paul Mellon. Maas's son, Rupert, believes that Mellon's support enabled his father to finance the Victorian picture exhibitions. Rupert went to work at the gallery with some reluctance in 1982, recounting how, when 'I heard how much my father owed the banks ... I realised that I had to stay'.[14]

The main competition for Maas was Julian Hartnoll, who started his career in 1962 at Newman and Sons, the epitome of Victorian genre taste. Hartnoll left three years later and, after a brief partnership with Giles Eyre, went on his own. He concentrated on a small stock of high-quality paintings, largely pre-Raphaelite. Hartnoll's stellar client list included pop stars and discriminating collectors such as Fred Koch and Isabel Goldsmith. Koch, who kept Victorian paintings prices high during the 1980s, had oil interests in the USA and formed an outstanding collection of the byways of Victorian art, similar to how Paul Mellon was collecting eighteenth-century British art, with less emphasis on the big names. In 1986, Koch announced that he wanted to give London its first museum of Victorian and nineteenth-century art, to be housed at St John's Lodge in

* One of those who saw it with Maas was Sir David Scott, who said, 'I remember it well.' Maas, who knew it had been lost for over sixty years, was incredulous. 'Oh yes,' said Scott, who was born in 1887. 'I saw it at the RA in 1897.' Nicolas Barker obituary in *The Independent*.

Regent's Park. Like so many before him, including Calouste Gulbenkian and Peggy Guggenheim, he found opening a museum in London more difficult than he envisaged and gave up.* Hartnoll also advised Joseph Setton, who formed a remarkable collection of pre-Raphaelite paintings in Paris, sold after his death in 1984 by both Sotheby's and Christie's. As Wood lamented, 'his tragic death, flying his own plane in the south of France, was a terrible blow to the Victorian revival'.[15]

Where the dealers went, the auction houses followed. In Christopher Wood's view, things began to move at auction with Millais's celebrated *Portrait of John Ruskin standing on the rocks at Glenfinlas*, which made a record 25,000 guineas at Christie's in 1965. When Wood suggested holding separate sales of Victorian paintings, he did not find support from Patrick Lindsay and Guy Hannen, his bosses in the paintings department. Going above their heads to the chairman, Peter Chance, the first sale was held on 11 July 1968, with a successful total of £74,000. Although the Atkinson Grimshaws sold for a few thousand pounds, most things sold for less.† One of the newspapers reported the sale with the headline 'Granny's Taste is Tops'.[16] Shortly afterwards, Wood published his *Dictionary of Victorian Artists* (1971), a useful book allowing dealers to look things up to make catalogue notes. The bibliography, with its list of periodicals, was especially helpful.

Sotheby's followed Christie's by instituting Victorian sales in 1970. Wood describes his opposite number, Peter Nahum, as 'a zealous and dynamic convert to the Victorian cause'.[17] Nahum – who could be astonishingly direct – was never short

* St John's Lodge had hitherto been part of Bedford College. Although they had been allowed to do all kinds of damage to the house, one of the reasons Koch was turned down was that it was perceived that he was going to spoil the house.

† Julian Hartnoll told the author that it was Colin Tennant who greatly stimulated the price rise in Grimshaw.

of superlatives and invented a new category: 'Highly Important Victorian Paintings', for which he produced the first all-colour auction catalogues of Victorian art. Wood usually inspected Sotheby's sales with Julian Hartnoll and Joseph Setton, and they would have dinner afterwards. When he left Christie's he became one of the leading dealers in the field.

Sotheby's main contribution to the Victorian revival was the establishment of 'Sotheby's Belgravia', an auction house dedicated to Victoriana and more besides. Based in the elegant, columned Pantechnicon in Motcomb Street, Belgravia, the saleroom promoted new markets, which the Sotheby's director Hilary Kay called 'the environment of new possibilities where groups of mavericks who liked novel areas were given their head'.[18] The brainchild of a young member of staff, Howard Ricketts, Sotheby's Belgravia was born out of Peter Wilson's attitude of 'Get on with it', and as long as things were running smoothly, he was not inclined to interfere. The saleroom had a transformative effect on the landscape of nineteenth-century art, giving self-respect to dealers in the area by treating their objects as art, as opposed to ornamental furnishings.

Sotheby's Belgravia was arguably a natural development because Bond Street had run out of space, but the new saleroom would assume a very different character. The key ingredient in its success was the group of young and enthusiastic specialists sent to run it, pioneers in their field and obsessed by their subject. Establishing the so-called 'collector's sales', these ranged from model steam locomotives, clockwork toys, cigarette cards, automata and anything that might be specialist collector material. Hilary Kay took over the Collector's Department aged only twenty, taking her first auction the following year. Although there was an entrepreneurial atmosphere at Belgravia, not every sale was successful and not every new market worked. The first dedicated sale of log burners, for instance, 'The Great Stove Event', was a notable success, but it sated the market, and the second sale, 'Decorative Heating Stoves', was a complete flop, with almost 95 per cent left unsold. More successful was Philippe Garner's first specialist nineteenth-century photographic sale in 1971, which firmly established the new category. The 'collectables' aspect of Belgravia attracted much attention and even a little satire. *Private Eye* advertised a sale of 'Highly Important Sausages', with matching carved oak-sticks, etc.

The initial problem was persuading Bond Street departments to release property for sales in Belgravia. The branding was different, with Belgravia's red-cover catalogues in a different format to Bond Street's green ones. One of the innovations was to illustrate all, or nearly all, of the items offered for sale, thus enhancing the burgeoning and scholarly re-evaluation of Victorian decorative arts. Another was to allow clients to pay their bill and collect the items during the sale, which today seems so obvious. Perhaps the most significant innovation came in the second season, when printed estimates were included in the catalogue for the first time in Britain. Printed estimates moved from there to Bond Street and then to auctioneers everywhere. As the *Financial Times* reported, 'The salerooms are progressing fast in the direction of demystification.'[19] Dealers were divided on the merits of this last initiative.

The Belgravia paintings department came into its own in 1973, with the sale of thirty-five paintings by Alma-Tadema from the collection of Alan Funt, just after the outbreak of the Yom Kippur War. In a time of economic uncertainty, the sale generated very high prices. Alongside Victoriana, Belgravia introduced one of its most successful new areas: rock'n'roll. The first sale was held in December 1981 and led to the Elton John sales in the next decade.

Belgravia was a fondly remembered saleroom of brief duration. Its days were numbered with Peter Spira's arrival at Sotheby's and fell victim to his charge to make the firm more profitable by stopping the duplication of cashiers, telephone operators, porters, etc. Although the outpost closed in 1985, its spirit survived through the Collector's Department, which Alfred Taubman, the new owner of Sotheby's, called 'the sports department' and where, in department store terms, the kids would go and get their first auction house experience. Many of the experts from the saleroom, notably Hilary Kay and David Battie, were involved in setting up *Antiques Road Show*, and something of the spirit of Belgravia still survives in that programme.

Andrew Lloyd Webber bought his first Victorian picture from Jeremy Maas, Millais's charcoal and gouache *Design for a Gothic Window* (1853). There is an oft-repeated story of the young Lloyd Webber seeing Leighton's *Flaming June* in a Fulham Road shop for £50, and attempting to borrow the sum from his high-minded grandmother, whose response was typical of her generation: 'I will not have Victorian junk in my flat.'[20] It was the success of his musical *Jesus Christ Superstar* (1969) that enabled Lloyd Webber to start collecting seriously at the age of twenty-two. In the course of the following thirty years, Lloyd Webber used several advisers, notably David Crewe-Read and, latterly, David Mason of MacConnal-Mason. It was clear from the beginning, however, that he knew what he wanted and had the ability, if not always the time, to acquire it. The Victorian paintings market rose with his success, and he was convincingly its dominant force for a generation. His auction apotheosis in buying Victorian art came in 2000, when his art foundation bought Waterhouse's *Saint Cecilia* (1895) at Christie's for a record £6.6 million. Lloyd Webber's exhibition at the Royal Academy in 2003 revealed him as the greatest collector of Victorian paintings since the nineteenth century.

The most flamboyant figure in the Victorian field was Roy

Miles, who brightened up the art world during the 1970s and 1980s. To some he wasn't a serious figure, but he made the art world more fun. Philip Hook described how, 'With a bouffant hairstyle which attested to his original calling as a successful society hairdresser, Miles opened lavish premises near St James's. The advent of any likely looking buyer into his gallery was met with a barrage of popping corks. If Frank Lloyd bullied his clients into buying art, Roy Miles did so by drowning them in champagne.'[21] Starting with old masters, he found it a difficult area to break into, and Peter Wilson advised him to concentrate on Victorian paintings, where there was less competition and more scope for his marketing skills.

Miles began by dealing from his flat in Eaton Place, but the Grosvenor Estate were not happy with him using it as a commercial premises and he moved near to Christie's at Duke Street, St James's. He met with hostility from the established dealers there and found he was being bid up in the saleroom. 'In those days,' he recalled, 'the art trade had a pecking order: this has now gone – today, only money is king.'[22] Undaunted by the snobbery surrounding his former profession, Miles enjoyed making a splash, dropping into country house sales by helicopter. For Christopher Wood, Miles's greatest talent was throwing parties, which he thought were the most glamorous in the art world: 'He always managed to conjure up some celebrities, and a photographer or two to record it all for the magazines. Half a page in the *Tatler* meant far more to Roy than any number of pages in *The Burlington Magazine*.'[23]

His grandest gesture of all arose from his desire to catch the Arab market, at a time when they had not hitherto shown much interest in buying art. As Miles recalled, 'I decided to visit Kuwait and take out a £1 million art exhibition. Bevis Hillier in *The Times* wrote it up beneath the headline "Oil To Arabs".'[24] A member of the ruling family asked him for a private view and, as he described:

> I walked our grand guest around the exhibition which covered the entire top floor of the hotel. As I did so, I identified each painting. 'This is a Thomas Gainsborough'; 'Here is a George Stubbs'; 'A fine Sir Alfred Munnings' and so on. The Prince did not say a word until we reached the door, when he turned and asked me 'How do you find time to paint all these pictures?' I quickly found $50 for the whisky.[25]

Leaving Duke Street, Roy Miles moved to Bruton Street, where he specialised in Russian art. His openings there were enlivened by handsome performers dressed as ornamental Cossacks, purporting to be from Russian state television doing interviews with Roy during the party.

David Messum's promotion of the Newlyn School was the bridge between the Victorian market and the emerging Modern British market in the 1970s. After an apprenticeship at Bonhams, Messum opened an antiques shop at Bourne End in Buckinghamshire in 1963. His first major paintings exhibition was *The Devonshire Scene* (1973), containing landscapes and marine subjects by artists like James Leakey, known as 'the Devonshire Wouvermans'. Messum began to specialise in the Newlyn School, holding the first exhibition in 1975. A decade later, he found himself sharing this market with Richard Green. Green, who has always been a major player in the nineteenth-century British paintings market, began to invest heavily in the Newlyn School, almost cornering the market at a regional sale held by Phillips on 13 November 1984. David Messum would henceforth often go half-shares with Green. As prices went shooting upwards, purchasing pictures in 'shares' became more frequent, particularly among the dealers at the top end of the market. However successful they appeared, most dealers had limited purchase funds, and buying in shares made it possible to find good paintings to show in their galleries.

Messum's dealership had a wobble in 1981, because, as he

accurately points out, 'Banks don't really believe in art dealers.'[26] NatWest foreclosed, and he had to close the gallery temporarily and remortgage his house. Messum came back with a new idea of approaching Modern British painting from its Victorian and Edwardian antecedents. He reopened in George Street at the back of Sotheby's in 1985, coining the term 'British Impressionism' and staging a series of exhibitions of British plein air painters, including Stanhope Forbes, John Lavery and the Scottish Colourists. In 1988, Messum held a major exhibition: *British Impressions: a collection of British Impressionist Paintings 1880–1940*. This was a savvy move, because by the end of the 1980s 'Mod Brit' had superseded Victorian as the market on the rise.

CHAPTER 19

The Londo Scene

'Anything and everything a chap can unload
Is sold off the barrow in Portobello Road.'

BEDKNOBS AND BROOMSTICKS (1971)

Beyond the West End, runners, street markets and alternative auction houses thrived. Philip Hook described how, 'Beneath those highest echelons flourished a secondary stratum of wheeler-dealers with smaller West End premises, men whose telephone lines were linked more regularly to their bookmakers than to the Louvre or the National Gallery. And beneath those again seethed a whole web of "runners", people who made their money from combing country sale rooms and emerged every other week in the West End trying to offload their trophies on the London trade.'[1] This formed part of the network through which a piece of furniture or porcelain might be passed through five or six hands before arriving at Mallett's Bond Street showroom, or one of the specialist shops in Kensington Church Street.

Until the internet made their trade redundant, there were dozens of runners who would arrive in London every week to show their goods to their favoured dealer or auction house. They had no premises other than their car or van from which to sell their wares on the pavement outside the gallery. The runner had, in the words of Jeremy Cooper, 'no overheads, no stock, no staff, no tax'.[2] They dealt only with the trade. Below them, at the bottom of the pack, were the 'knockers', some of whom were fences and conmen, going from door to door searching for bargains. They achieved a certain notoriety thanks to Roald Dahl's story 'Parson's Pleasure' (1977). These rogues had a typical ploy: the woodworm trick. While examining a bit of furniture or a mirror, they emptied a tiny amount of sawdust from a concealed matchbox. With a great sigh of disappointment, they would draw the owner's attention to the little pile of 'woodworm' and then come to the rescue: 'I'll take it away and deal with that for you.'

The centre of the world of runners was Brighton, and they were sometimes known as 'Brighton Runners'. Brighton was a fascinating place in the 1960s, an extraordinary mixture of thespians (Laurence Olivier lived there), gay grandees who liked the Regency backdrop and the gangland underworld Graham Greene had described in *Brighton Rock* (1938). The veteran auctioneer Dendy Easton recalls that it was full of characters called 'Bubbles' or 'Fingers' (he had lost most of them), who had an amazing instinct for objects and often came from families who had run fruit and vegetable barrows before going into antiques. The runners had great charm and were generally gamblers who would spend their money as soon as they made it.[3] Occasionally, they found very good pieces – a runner one day brought John Baskett a painting which turned out to be a Hogarth. The downside was often the recent history of the object. Danny Katz recalled a runner coming into his gallery and introducing himself as 'Big John': 'I've got something nice,' he announced and produced a huge Paul de Lamerie silver salver. 'Is it hot?' Danny asked gingerly, at which Big John frowned, 'Lukewarm', and Danny sent him away.[4]

Stephen Somerville recalls that the runners' network could be very positive. Sometime in the 1970s an old master painting was stolen from Colnaghi's in a hit and run raid at 10 a.m.

Within half an hour it was offered to a runner, who identified it as the kind of picture that Colnaghi's might like, and it was back with them by 11 a.m.[5] Runners were usually male, although John Baskett fondly remembered a Mrs Thrupp, who brought in pictures once a week. The most improbable runner was a psychiatric doctor from Geneva, Dr Ley, who touted pictures along Duke Street on approval from one dealer to sell to another on the same street. Philip Hook recalled 'the splendid Michael Fulda, whose fine eye for provincial discoveries was matched only by his parsimony. When his old van finally gave up the ghost, he found a more capacious vehicle at a knock-down price: a second-hand hearse. At the back door of Christie's, it was sometimes difficult to tell if there had been a bereavement or it was merely Fulda delivering pictures.'[6]

A cut above the rest, the Rich brothers from Marlow were, in Julian Hartnoll's opinion, 'extremely good second generation – rather academic – runners', although, as he added, 'one never asked where something came from'. Paul Rich had a gallery in St Christopher's Place that only dealt with the trade: 'You can buy pictures by the stack,' he would say to David Messum. Many of the runners came from the North, especially around Harrogate. The Coulter family were based in Tadcaster, Yorkshire. The father had a business smelting the gold leaf off frames (this was quite common after the war) and would send the unframed pictures to Bonhams. When one turned out to be a Stubbs and made £12,000, the Coulters took more interest in the paintings, and later opened a gallery in York.

The family-owned Bonhams was the runner's favourite secondary saleroom, partly because it was more accessible in Knightsbridge. With a history stretching back to the eighteenth century, the firm built its own saleroom on Montpelier Street, inaugurated in 1956. One of Bonhams main selling points was that it would do a complete house-clearing service, something the big auction houses could not possibly attempt. When David

Messum joined in the early 1960s, he found himself cataloguing up to 400 – mostly Victorian – pictures a week. When Dendy Easton walked into Bonhams in 1971 and asked for a job, he was immediately made a furniture porter. He soon transferred to pictures under Alex Meddowes, where a great deal of their supply came from two Northern families of runners, the Sutcliffes, and the Coulters, who were responsible for as many as eighty lots in any Bonhams picture sale. Runners were forced out by the rise in auction house thresholds and, above all, the internet. As Julian Hartnoll observed, 'Access to the many sites that list forthcoming auctions was a death sentence to the runners.'[7]

The porters at Bonhams assumed an extraordinary importance. They were mostly raffish public-school types who had not gone to university but were notoriously well-informed. They would frequently buy lots themselves if they were cheap, putting them back in a later sale when there was a bigger attendance. It was said that they used to double their weekly wages with these reselling activities which nobody thought was wrong in the 1970s.[8] As a porter, Easton recalls the hazard of being buttonholed by the dealer, Mr Dent – 'come 'ere boy' – with his repertoire of dirty stories which he would regale to anyone prepared to listen.

Bonhams was the first of the main London auction houses to have a female auctioneer. From the 1940s onwards, Helen Maddick, sister of Leonard Bonham, took sales. She was a commanding presence, but not as intimidating as the lady auctioneer at Robson Lowe in the 1950s, who interrupted her sale twice to tell a revered client to stop talking before eventually hitting him very hard on the nose with her auctioneer's hammer.[9] The doyenne of female auctioneers was Miss Dorothy Bagnell, who took cigarette-card sales at Caxton Hall, and despite doing 100 lots an hour, found time to enliven her performance with protestations like 'Oh dear, *another* parcel to send abroad.'[10]

Bargains could be found in this happy-go-lucky atmosphere.

Anthony Speelman happened to be in London in August 1966, a dead month in the art world, when most people packed up and went on holiday. Bonhams had just received a large collection of paintings from a deceased estate in Bournemouth for immediate sale. Speelman was surprised to see two partners of Agnew's: Colin Agnew and Dick Kingzett, and correctly guessed that the reason for their presence was a very small portrait of a monk holding a scroll of paper. Less dirty than most of the other paintings, it was evidently a fifteenth-century Flemish work of exquisite quality: 'The name that came to mind as I looked at it was the Master of Flémalle. I looked around the room to see if anyone was watching me, but Colin and Dickie were not visible.'[11] Leaving the saleroom, Speelman could not resist another look, and this time the Agnew's directors spotted him. As he left, they followed him to Bond Street: 'On the way there, our conversation was rather like two boxers in the first round, neither one wanting to reveal his strong points.' Finally, they agreed to buy the picture in shares with a bidding limit of £20,000. In the event, they bought it for £1,800 and sold it to the National Gallery for £60,000.*

Post-war London had dozens of minor auction houses, including Roger, Thomas and Chapman near Cromwell Road and Robinson and Foster near King Street opposite Christie's, both of which have disappeared. One of the most characterful local auction houses was Forrests in Leytonstone, which would hold huge sales of 'Antiques and Objets d'Art' every six weeks – 'Now this is rather exquisite', the auctioneer would call about a pair of nineteenth-century seascapes.[12] Beneath the arches of a viaduct over which trains rumbled, the auction room was a large central space with several arched smaller rooms for viewing, offices and a tea and sausage roll counter.

Many salerooms, including Druce's of Baker Street, Knight

* The picture has since been demoted to the workshop of Robert Campin.

Frank and Rutley (household goods division) and the Army and Navy, closed during the 1960s and 1970s. This left Harrods as a survivor of the days of disposals of properties from their large depositories, until that closed too. Harrods' sales were held in their warehouse on the Thames, which achieved fame once a year during the Oxford and Cambridge Boat Race. In 1960, Basil Taylor, the Stubbs expert, noticed *Queen Charlotte's Zebra* coming up at Harrods in a sale of furniture and household goods, and alerted Paul Mellon. The collector naturally hoped he might get a bargain, but in the way of such things, so did everybody else, and he ended up paying £20,000. Harrods auctions fizzled out in the mid-1970s, and by 1979 the big four auction houses dominated the scene with a few survivors like Lots Road. One of the better known was Debenham Coe in Brompton Road. Bill Brooks, the managing director, was a flamboyant auctioneer, describing Chinese vases as 'full of occidental promise' and their sales were 'grand entertainment, as Brooks alternatively wooed and browbeat his audience into paying decent prices'.[13]

In 1975, when Christie's was looking for a secondary saleroom to combat Sotheby's Belgravia and, above all, relieve pressure on King Street, they acquired Debenham Coe. Although different in intention from Belgravia, Christie's South Kensington, always known as 'South Ken', sold many of the same kind of items: costumes, embroidery and Victorian photographs. Bill Brooks became the first joint managing director with Paul Whitfield. Brooks thoroughly enjoyed his new role, losing none of his ebullience as he threatened the audience with turning on the fire extinguishers if they didn't wake up and start bidding. He had a rolling, musical voice which gave a Victorian air to proceedings. South Ken became a much-loved fixture on the dealers' itinerary, and the most successful attempt by either Sotheby's or Christie's to run a secondary saleroom in London.

If Bonhams (until the end of the century) occupied the number four position, Phillips, Son & Neale was third in the pecking order, striving to take on Sotheby's and Christie's, usually with disappointing results. The firm rose dramatically in the 1970s, from a turnover of £4 million at the beginning of the decade to over £27 million towards the end. Running three salerooms in London, the headquarters were just off New Bond Street in Blenheim Street. The chairman, Christopher Weston, was all-powerful and something of a dynamo, believing in efficiency and fast payout. Coming to the firm aged thirteen during his school holidays from Lancing, Weston became chairman in 1972 at the age of thirty-five. He had little charm and no time for small talk or glamorous parties. When Richard Knight started at Phillips in 1969, he found Weston 'an aggressive little man', a businessman rather than an art lover.

Weston decided that taking over provincial auctioneers would be the answer to expansion, and he did it much more successfully than Sotheby's. At its height, Phillips had a network of thirty-two regional offices and salerooms.* The firm had several excellent specialist departments, including furniture, ceramics, collectables and jewellery, and it pioneered sporting memorabilia. It was when Weston tried to break into the New York market and impressionism during the 1980s that things went wrong. He made another stab at this market in the mid-1990s, this time doing what the firm knew best, working with North American regional salerooms. After a health scare in 1998, Weston sold the business to his brother, allegedly for £14 million.[14]

* In the early 1990s Weston once rang his MD, Richard Madley, to say, 'I just bought sixteen more salerooms', referring to the network belonging to the Prudential.

The art world has always enjoyed lunch, and it continued to do so long after it went out of fashion in the City. Dick Kingzett's obituary underlined how, 'In London, lunch remained an important, unhurried part of the working day. Dick Kingzett impressed upon his sons that while Cognac might well militate against a productive afternoon, Calvados never did anyone any harm.'[15] Everyone had their favourite watering hole, depending on their preference for speed, discretion, grandeur, the need to be seen or simply proximity. At one end of the scale in St James's there was White's or Wilton's, somewhere in the middle were the numerous Italian family-owned restaurants, and then at the markets there were the various snack bars, where dealers would congregate and show each other objects.

Wilton's in Jermyn Street was the choice for the big deal. David Somerset had a booth reserved there, and it was Julius Weitzner's favourite. Not every lunch at Wilton's had the desired effect, however. Due to meet Somerset there on one occasion, Baron Thyssen was early and, as it was pouring with rain, he rang Anthony Speelman's doorbell for sanctuary. There he saw a beautiful Segers painting and could not resist buying it. Two hours later, Speelman's telephone rang, 'It was David. He was apoplectic: "I can't believe it. That bloody man goes into your office to shelter from the rain and spends a quarter of a million pounds and then tells me he has no money left to buy the pictures I showed him after lunch." I, of course, roared with laughter and in the end, David did see the funny side.'[16]

White's was sometimes referred to as 'The Art Dealers' Arms', but in truth only a small number of Christie's directors and a few dealers were members. Hugh Leggatt could be seen most days lunching his two sons at a conspicuous table near the entrance, but the two dealers who probably benefited most from their regular attendance were Simon Dickinson and his partner, Dave Ker. When they were not lunching at White's, the Christie's grandees, Patrick Lindsay, Guy Hannen and

Charles Hindlip, would patronise Wheeler's in Saint James's, where the catchphrase was 'one bottle of Muscadet doesn't do for three people'. More generally popular than White's among the grander trade was Brooks's, where Geoffrey Agnew and Jack Baer held sway. Christie's directors like Hugh Roberts, Noël Annesley, Hugo Morley-Fletcher and, more rarely, Tim Llewellyn and David Moore-Gwyn from Sotheby's could be found there.

The contemporary world, its artists and hip hangers-on would gather in Soho at the Colony Room (opened by Muriel Belcher at 41 Dean Street in 1948, and run by Ian Board after her death in 1979) or the Groucho Club, and to a lesser extent the Chelsea Arts Club. The Soho world around the Colony Room is wonderfully described in Daniel Farson's book *The Gilded Gutter Life of Francis Bacon* (1993). It was, however, over lunch at another Soho institution, La Trattoria Terrazza, always known as the Trat, that Bacon, Farson and the gallerist James Birch had lunch in 1986, when the latter put the idea of a Moscow exhibition to Bacon that came to fruition two years later.

If Christie's directors took their grander clients to Wilton's, Overton's or Prunier's, for the rest of the staff it was luncheon vouchers and a scramble to get a seat at The Bonbonnière, an Italian cafe in Duke Street. Brian Sewell describes it as:

> used by the slaves of the art market, the lowest forms of saleroom life: packers of parcels and collectors of purchases, the commission agents who never betrayed the skeletons in their employers' cupboards, and every itinerant workman unfortunate enough to have employment in the neighbourhood. At the 'Bonbon' the rule was to share tables and shuffle seats, order quickly, pay and go, for only by packing us in and throwing us out, could the ever-smiling Italian family that ran it keep their prices down to a level that made the Luncheon Voucher a viable bill of exchange.[17]

Several dealers had their own reserved table, and Andrew Ciechanowiecki's former spot at Frank's (now Franco's) in Jermyn Street is today marked by a plaque. Frank's was also the favourite place of Ridley Leadbeater – the head of Christie's front counter – to celebrate when he had pocketed a good tip. Leslie Waddington had his own table at the Caprice, while Johnny van Haeften preferred Greens. Van Haeften remarked on the importance of lunching clients: 'art is a very personal business. If I am to collect as a hobby it must be made enjoyable and entertaining.'[18] Dealers tended to divide between those who hosted clients and those who went to local bistros with their colleagues. Richard Green bucks the trend of sociability and enjoyed the pies at The Guinea in Bruton Place instead.

Overall, the family-run Italian restaurants were the favourites. Towards the northern end of Bond Street, The Chalet in Maddox Street was patronised by staff at the Fine Art Society, Phillip's and Sotheby's. It was at The Chalet that Elizabeth Wilson had recognised the photograph of De Vries's *Dancing Faun*. Kasmin favoured the Gay Lord restaurant in Albemarle Street. However, when he wanted to take David Hockney out to lunch early in their relationship at the Marlborough Gallery, he recalls, 'I was thinking in terms of Wilton's, but Hockney pre-empted me being a vegetarian, and knew about this place off Leicester Square called Vega, where the bill came to one and ninepence. David saw me looking at the bill and said, "Oh Mr Kasmin, is it too much?". "No" I said, "it's certainly not, but it's very difficult to explain a bill like this to anybody".'[19]

At Le Petit Café in Stafford Street, Anthony d'Offay would lunch with the bookseller John Saumarez-Smith and the publisher John Murray. Owned by two Italians, the star was the cockney waitress, Elsie. It was 'an inexpensive sandwich shop that had five tables upstairs and delicious simple food. You took your own wine. Elsie ran it with a combination of no-nonsense authority and motherly affection for her regulars. If she did not like the look or manners of a potential guest, they were not given a table.'[20] When Richard Day took Paul Mellon out to lunch, he realised that Claridge's would not be amusing for him, and took him to Le Petit Café instead: 'as we climbed the creaking wooden staircase to the first floor with the tables, we found Sir Oliver Millar, the Surveyor of the Queen's pictures, and the Agnew's director, Evelyn Joll, already lunching. Paul greeted them and we sat down. "What's you gents 'aving?", Elsie shouted, to the delight of Mellon.'[21]

Some dealers had their own dining room, notably John Partridge and the Norton family at S. J. Phillips, but the most entertaining was done in the boardrooms of Sotheby's and Christie's. Up until 1974, when Jo Floyd took over as chairman, the art trade was never invited to lunch in the boardroom at Christie's. The first to be thus honoured were David Somerset and Anthony Speelman, hosted by Patrick Lindsay, who 'liberally dispensed his views of the art trade which were not always very favourable'.[22] After lunch, Speelman and Somerset were walking back to their respective offices when Somerset turned and said, 'I now know how it feels to be the Keeper invited to the Big House for lunch', and they both roared with laughter.[23]

Too busy to enjoy lunch, Peter Wilson tended to rush to the rather characterless Westbury Hotel because it was next door, quiet and – owing to its dullness – discreet. Shortly after Alfred Taubman acquired Sotheby's, the Queen Mother let it be known that she would be amused to meet him. Lunch was duly arranged in the Sotheby's boardroom. Taubman was surprisingly nervous, but she put him at his ease and things went very well until the coffee. A new visitors book was presented to the Queen Mother for signature. As she wrote, Taubman – visibly relieved that lunch had gone so smoothly – was telling a story, and in an expansive gesture, knocked his coffee all over the newly signed page. Unfazed, she picked up the blotting paper, saying: 'Did you know it was the English contribution to the

Florence floods to provide blotting paper.' Taubman spent the rest of the afternoon intoning, 'What a dame!'

One place the art trade would congregate was at St George's Gallery Books in Duke Street run by Agatha Sadler, one of the most universally admired figures in the art world. Sadler was gossipy and sometimes waspish, and her gallery became something of a hub, and a vital supplier of rare art history books and catalogues. To Brian Sewell, 'No art historian of my generation, European or American, can have pursued his profession without dependence on Agatha and her bookshop.'[24] The antiquities dealer Robin Symes threw a grand party for her at Bibendum when she retired. Her place as the premier art trade book dealer was taken by Thomas Heneage, also in Duke Street.

Primarily a way of meeting new customers, fairs appear throughout this book, many of them specialist, but where they intersected was at the Grosvenor House Fair, the giant of them all while it lasted. Initially held around major auctions, fairs began to dominate, and today it is the other way round. As the power of the auction houses grew, fairs became the dealers' way of fighting back. They were useful to put pressure on prevaricating clients to make up their minds before something was taken to be exhibited. Perhaps fairs' greatest influence was the creation of vetting committees that raised standards across the trade.

Grosvenor House was founded in 1934 to beat the Depression and lasted for seventy-five years. It was a very English market in the early days, selling brown furniture, silver and jewellery, and from the 1950s it accepted paintings, which gradually came to dominate. The Grosvenor House British Antique Dealers Association Fine Art Fair, to give it its full name, engendered a degree of cynicism within the trade. As Anthony Speelman explained:

> It was as though the dealers had brought everything that they had been unable to sell in the past years for reason of condition or poor quality, hoping to unload them on fresh clients who needed to cover their walls while furnishing their houses. One stand, owned by a partnership of two dealers, had twenty paintings on it, by the time the vetting committee had finished it was left with six. We watched the two dealers ferrying the paintings away, rather like stretcher bearers during the war. Of course, they then had to go around London to borrow enough paintings to cover their walls.[25]

There was a camaraderie among the exhibitors and a certain amount of mischief. One day, an impish impulse seized Speelman when he saw Richard Green's stand being put up. The sign advertising his name was waiting to be put in place and he removed the letter 'N' from Green and replaced it with the 'D' from Richard, who was not amused. Grosvenor House was probably at its zenith in the 1980s, when the quality was at its highest, with eighty British dealers exhibiting. Mrs Thatcher opened it in 1988, and the following year was said to be the best year ever, with everything sold, according to rumour. Designed for American visitors who were gradually diminishing, the seeds of Grosvenor House's destruction were sown by the appearance of the new international fair at Maastricht. Its closure in 2009 was the result of a failure to adapt.

Maastricht Fair, later TEFAF, opened in the mid-1970s as the brainchild of Robert Noortman, an energetic and successful dealer in Dutch old masters who hailed from the town. Creating Maastricht was an extraordinary act of confidence. Although in the middle of Western Europe, it was an awkward place to get to, but the fair flourished from the beginning.*

* Grosvenor House would be replaced by Masterpiece in 2010, beyond the dateline of this book.

There was a slight emphasis on Dutch and Flemish pictures, and by the 1980s almost 100 dealers (about twenty from London) were exhibiting 2,500 paintings worth £150 million.* It struck a blow against London, but the London dealers who went there prospered and met new clients, especially from Germany, the Netherlands and Belgium. Johnny van Haeften, the quintessential London dealer, first exhibited in 1978, and he became joint chairman shortly thereafter.

Nowhere did the art trade and the public more enjoyably mingle than at Portobello Market. For a great many Londoners, visiting the market became a Saturday morning ritual. It was also a training school, where many who went on to have distinguished careers had their first stall.

On Saturdays from 6 a.m. to 9 a.m. it tended to be a trade market, and the public would arrive between 7 a.m. and 10 a.m. There were three levels: the barrows, the arcades (where the most interesting material was to be found) and the flanking shops. Nobody dared miss Portobello, a vital link in the 'spider's web' between country and town dealers. It was the place to find specialist material, especially glass, tribal art and porcelain. One stallholder, Kathleen Skin, discovered a niche in buttons with insignias. She remembered Portobello during the 1960s with people dressing up, squeezebox bands and one stallholder with performing monkeys. A lifestyle choice, as one trader put it: 'Antique dealing is a very independent way of life … It's a business for autodidacts, if you like: people who have educated themselves.'[26]

Portobello attracted grand *marchants amateurs* such as Sir

* The fair prospered and grew rapidly to 275 dealers arranged over 'streets' with names such as Unter den Linden or Champs Elysées or Trafalgar Square.

Martin Wilson Bt, Peter Wilson's elder brother. Inheriting the contents of his family home, Eshton Hall in Yorkshire, these became the principal items for sale in the shop he set up with his butler, Leonard. If somebody attempted to bargain, Sir Martin would say, 'I will have to ask Leonard about that.'[27] Portobello attracted literary and musical attention, from Muriel Spark to the Sex Pistols. Paddington Bear was a regular visitor, and Tom Courtenay in *Otley* played a light-fingered Portobello antiques dealer pretending to be a spy. Cat Stevens found himself 'walking down Portobello Road for miles' seeing 'cuckoo clocks and plastic socks / lampshades of old antique leather / nothing looks weird, not even a beard / or the boots made out of feathers'.[28]

If Portobello was a tourist Mecca, Bermondsey Market was more of a trade affair. Everything was done initially by torch-light, and there were often stolen goods. Big lorries would arrive at 2 a.m. and park in a circle like wagons in a Western. At this time in the morning, the dealing was entirely trade-to-trade. Punters made offers as the dealers emptied their boxes and battered suitcases. By 4 a.m. the local council had put out long trestle tables, and it was a different story. To keep up the dealer's strength several well-known cafés served gargantuan breakfasts. At Bermondsey, purchasers in good faith got good title if the item was sold openly between dawn and dusk, i.e. stolen goods were not liable to be returned, reflecting the ancient law of *marché ouvert*, abolished only in 1995.* At the

* A turning point came when, in March 1993, a gentleman brought into Sotheby's two dustbin bags containing two portraits, one by Reynolds and one by Gainsborough. It was quickly established that they came from Lincoln's Inn, from where they had been stolen. Because they had been bought in a *marché ouvert*, the consignor had good title and could sell them. The publicity certainly helped the abolition of *marché ouvert* in January 1995.

lower end, the antique trade was a cash economy, lacking receipts or documentation.

The king of the London antique centres was Bennie Gray, who established his first antique centre in Barrett Street in 1964. This was followed by the Antique Hypermarket (1968) in Kensington; Antiquarius in Chelsea (1970); Alfies (1976) in North London; and the eponymous Gray's Antique Market in Davies Street (1977). Gray spotted the gap in the market for art nouveau and art deco:

> As all this began to overlap with the exploding pop culture, it created a gaping hole in the structure of the antique trade ... Most of them wanted something more permanent than a street market but couldn't afford to rent a formal shop in a decent location, so I set about looking for a cheap building in central London big enough to accommodate a number of small antique shops. It didn't take me long to find an old printing works on a short lease in Barrett Street, just around the corner from where I lived in Manchester Square. It was ideal – seconds from Oxford Street, gritty, spacious and very cheap. I signed the lease, cleared out the old printing machinery and built 27 stalls like the ones in the Portobello Road. I called it the Antique Supermarket and at £5 a week there was no shortage of takers.[29]

Kathleen Skin graduated from Portobello to Gray's market, where she recalled the dealers all looking after each other's stalls at lunchtime, relishing the day when one of them made a big sale and they would all go off to Claridge's for lunch.[30]

When the IRA blew up Nelson's Column in Dublin in 1966, Bennie Gray caused a sensation. Amazingly, Nelson's head had survived intact, but had been stolen. Gray hot footed to Dublin, where he bought every bar crawler a drink until, on the third day, he found the 'guardians' of the missing head. Bringing it to

London, he put it on show in the antique supermarket and drew the crowds. As Gray told the author, 'Overnight we had become a cool destination – no small achievement for a bunch of small-time antique dealers at the very peak of swinging London.'[31]

Chapter 20

The Zenith of Sotheby's and Christie's in London

'The Sunflowers was the apex of the story.'[1]

Desmond Corcoran

The period between 1977 and 1995 saw the summit of Sotheby's and Christie's London salerooms' importance before their gradual repositioning into second place behind New York. It saw the final triumphs of Peter Wilson with the Mentmore and von Hirsch sales before his retirement and the turmoil that followed until Alfred Taubman acquired Sotheby's and turned it into an American company. Christie's, which had been running at roughly half Sotheby's turnover since the 1950s, dramatically caught up during the 1980s. The two firms had been differentiated by their core clients: Sotheby's impressionist collectors and Christie's grand British clients selling old master paintings. This distinction dissolved as the two firms fought ferociously to win the other's ground and Sotheby's broke into the country house market while Christie's began to win great impressionist sales. The battleground for market share would be the modern sales in New York, featuring in particular the works of Van Gogh and Picasso, which broke records during the 1980s and 1990s. The sale of one of the seven versions of Van Gogh's *Sunflowers* in 1987 was the last time the world record for the most expensive painting would be held by a London saleroom.

When Christie's opened their saleroom in New York in 1977, they charged a buyer's premium, while Sotheby's New York had resisted the idea, believing this would bring consigners' good will. David Bathurst, Christie's charismatic head of office, explained: 'The initial issue was the premium. We had it and the opposition didn't ... the premium allowed us to lower the vendor commission and that's all the seller was concerned about.'[2] Two years after Bathurst's appointment, the 'turning point' came when he secured the sale of ten pictures of the highest quality, including two Van Goghs, a Cézanne and a Picasso, from Henry Ford's collection. The sale, on 13 May 1980, made $18.30 million (£8 million in the money of the time), well over the estimate of $10 million. As one commentator observed, 'Sotheby's had grown stale, jaded by their unchallenged success in the United States, and were wrong-footed by the young and enthusiastic team that Christie's unleashed on America, led by David Bathurst.'[3]

Bathurst was soon, however, to suffer from his own hubris. His career came to an end after an impressionist sale in New York in 1981, when only one of eight paintings found a buyer. Faced with market collapse, Bathurst, an instinctive risk-taker, lied to the press, announcing that three paintings had been sold, rather than one. This lie might never have surfaced except that the seller, Dimitri Jodidio, angered by the bad result, sued Christie's for negligence in the US Supreme Court. Although the claim was rejected, Bathurst's deception emerged, and for once it was Christie's, not Sotheby's, in the eye of the storm. Bathurst resigned, leaving behind an excellent successor, Christopher Burge, who continued piling pressure on Sotheby's in New York over the next decade.

If Christie's were successfully challenging Sotheby's in New York, the latter would pull a cracker from Christie's traditional terrain in Britain. Following the death of the 6th Earl of Rosebery, the contents of Mentmore Towers in Buckinghamshire provided the most spectacular house sale in living memory. The contents, which had a Rothschild provenance, was especially strong on eighteenth-century French decorative arts. The sale in May 1977 lasted ten days and seemed like the end of an era of grand living. Despite the opulence, cataloguing conditions were austere. As James Miller recalls:

> Preparations for Mentmore started in late October, with a view to holding the sale the following May. We would meet every morning at 5.30am at Bond Street with whichever expert we wanted to come with us. A van took us down. We always ate breakfast at the Happy Eater café on the way. Because electric lighting at Mentmore was so dim we could only catalogue between 8.30 a.m. and 4.30 p.m. in winter. Chadwick, the family butler, would run a hot bath every morning at 11.00. We could all go and put our feet into that bath because there was no heating in the house. It got the circulation going again. Sherry was served at midday.[4]

The conservation furore to 'Save Mentmore' gave vast publicity to the Sotheby's sale, a fashionable event with helicopters flying in and out. Buyers competed as much for the contents of the maids' rooms as the drawing room. David Carritt spotted the 'sleeper' in the sale, a painting described as *Toilette of Venus* and traditionally attributed to Carle van Loo. From a careful reading of the Fragonard catalogue raisonné, he remembered an untraced work of *Psyche Showing her Sisters her Gifts from Cupid*, clearly the real subject of the Mentmore painting. He secured the work for £8,000 and sold it to the London National Gallery as the lost Fragonard for £495,000 a year later. The Mentmore sale raised £6.39 million, pulling the art market out of the 1970s slump.

The year after Mentmore came Peter Wilson's apotheosis with the sale of the treasures of the Basel collector Robert von Hirsch. Wilson had waited all his life for this sale. He had met the family when he had gone to Germany after leaving school to learn German, and had become friendly with the collector's stepson. The sale represented everything he stood for: great European collectors, connoisseurship, the highest-quality medieval works of art and a very desirable collection of impressionists. The von Hirsch sale was the most important collection brought into London in modern times and is, alongside Van Gogh's *Sunflowers*, the zenith of this story. Nothing like it for quality and range had come onto the market since the war. Above all, it was the medieval treasures that enthralled the world.

Von Hirsch had left Germany in 1933, having already transferred his business interests to Basel. He was allowed to leave with his art collection on condition that he gave Lucas Cranach's *Judgement of Paris* to the German state.* Visitors to his Swiss house recall that each room was devoted to a different collecting area: impressionists in the drawing room were hung in two rows, twentieth-century works were displayed in the dining room, while the heart of the collection, the medieval works of art, were placed in the library. Part of the sale's appeal was the small scale of everything. Many asked why von Hirsch did not want to keep his collection together. The answer was that it was very much a private collection formed for private pleasure and, as his friend Jürg Wille wrote in his introduction to the catalogues, von Hirsch 'believed in the free circulation of works of art and wanted his collection to be distributed on the market'.[5] As the collector put it, 'I want people to love and fight for my things just the way I did.' The sale caused an international sensation within the museum and collector world, and Wilson sent

* Restituted after the war and now bequeathed to the Basel Museum.

a selection of the items on tour to Zurich, Frankfurt and to the Royal Academy in London beforehand.

The von Hirsch sale's 674 lots* were spread over four catalogues: old master paintings and drawings (including a very rare Dürer landscape), medieval and renaissance works of art, Continental furniture and textiles, and finally a catalogue of impressionist and modern paintings, drawings and sculpture. A Swiss travel agency arranged special package tours to London for the auction. The German museums would be the main winners. The federal state made £10.8 million available to a museum consortium on the basis that this was a last opportunity to acquire such important medieval objects. Rainer Zietz and Sir Geoffrey Agnew bid on their behalf. The British Rail Pension Fund was represented by Giuseppe Eskenazi and, as previously mentioned, bought the Gloucester Candlestick for £550,000. The sale took six days during June 1978, and soared above its pre-sale estimate of £8 million to make £18.5 million.

By 1980, Wilson was diabetic and in poor health. Following the great triumphs of Mentmore and von Hirsch, he decided to retire to his house near Grasse, hoping he could still pull the company strings through his successor, the Earl of Westmorland. As Geraldine Norman realised, 'It wasn't going to be a real retirement. It was a miscalculation because he couldn't run such an enormous empire from anywhere save his office in Bond Street.'[6] Without Wilson, corporate rivalries and hatreds emerged under Westmorland's benign but weak chairmanship. The managing director, Graham Llewellyn, disliked the vice-chairman, Peregrine Pollen, who disliked the finance director, Peter Spira, and so it went on. Gordon Brunton was brought in from the Thomson newspaper empire to sort out the mess. He fixed on the cleverest and least controversial of the directors,

* The sale went up to lot 877, which is misleading, as there were only 674 lots in the sale.

Julian Thompson, head of the Chinese department, to replace Westmorland. Brunton wisely observed that, 'A company is at its most vulnerable when it is putting its house in order but has not managed to get those improvements reflected in its price on the stock exchange.'[7] He was right, and Sotheby's was potentially exposed to a hostile takeover.

Into this unstable situation two predators appeared. A pair of American entrepreneurs, Marshall Cogan and Stephen Swid, made a bid of £5.20 a share in April 1983, with a proposal that Lord Harlech, ex-ambassador to Washington, become chairman. The Sotheby's home team did not like the look of Cogan and Swid, dubbed 'toboggan and skid', and felt there was no synergy between their business and Sotheby's. It was Julian Thompson who told them: 'It will not work, you are simply not our kind of Americans.'[8] Sotheby's was spared by a referral to the Monopolies Commission, on the grounds that the company was uniquely British and such a takeover was not in 'the public interest'. This brought derision from one Conservative MP, who snorted, 'How can Sotheby's claim to be part of the national heritage when they have made so much money selling off bits of it?'[9]

In a chance conversation between Lord Westmorland and a City figure, David Metcalfe, Westmorland mentioned Sotheby's was looking for a 'white knight' and Metcalfe suggested A. Alfred Taubman. A Detroit property developer (and pioneer of the shopping mall), Taubman was an art collector married to Judy, an ex-Christie's employee. The Taubman takeover was welcomed by the Sotheby's board and went ahead. Taubman was a generally benign main shareholder until things went wrong at the end of the millennium with the New York collusion case. A large, smiling man, at his best wise with a kindly manner, Taubman would become more intimidating over the years. He created a Holdings Board with some of the richest people in America, including Henry Ford, Max Fisher and Leslie Wexner, alongside Baron Heini Thyssen from Europe. There was initially nothing very radical about his stewardship of Sotheby's, aside from an increasing emphasis on client services and searching out buyers.* Hitherto, both Sotheby's and Christie's had devoted 90 per cent of their energies on finding property for sale, believing that good material would sell itself.

George Bailey, a later European managing director, thought that 'the first years of Alfred Taubman were the best years ever'.[10] The years 1984 to 1989 saw extraordinary growth akin to the 1960s, with an increase in turnover from £500 million to £1.8 billion. Some of this can be put down to the rise of Japanese buyers, who dominated the market for items fetching over $1 million. In 1984, Sotheby's London achieved a world record, selling Lord Clark's Turner, *Folkestone*, for £7.37 million to the Thomson family of Canada. However, Taubman's main interest in the London office was as an architect *manqué*, trying to make sense of the rabbit warren of offices. He would spend all day with the house architect, Tony Marsh, poring over plans for placing a café in the middle of the building, or enlarging the main galleries.

Taubman's choice of chief executive, Michael Ainslie, was a good-looking, preppy figure based in New York who tried to give the London office the illusion of still managing Europe. This fiction got thinner until Dede Brooks, who took over his position in 1993, dispensed with it altogether. Ainslie immediately wanted to raise the threshold for accepting property for sale to $1,000. This was fine for New York, but presented problems for the London office, which had to deal with property from country house clients and needed to service house sales

* In 1984, Alfred Taubman gave a famous interview to the *Wall Street Journal* in which he claimed that, 'Selling art has much in common with selling root beer. People don't need root beer and they don't need to buy a painting either. We provide them with a sense that it would give them a happier experience.'

and valuations. It was especially difficult for departments with low lot values like books and porcelain. The argument over the lower and middle markets would dominate Sotheby's and Christie's internal debates for the next thirty years.

Taubman appointed the Earl of Gowrie as the new chairman of Europe and promoted Tim Llewellyn from the Old Master Department to be the new European managing director. Grey Gowrie was an imaginative choice, a former arts minister under Margaret Thatcher, a poet and sometime dealer with Thomas Gibson (in best establishment tradition, his old fag at Eton). Gowrie had brilliance, charm and, above all, eloquence. Sotheby's still had something of an inferiority complex about the UK, with ducal art collections like Chatsworth still firmly wedded to Christie's. In marketing terms, Gowrie's appointment rescued the firm from the rackety reputation of the Wilson years and highlighted Christie's slightly stuffy image. Gowrie would compare Sotheby's to the strap line of the car rental firm, Avis: 'we try harder'. There was the lingering suspicion, however, that he treated the job as if he were a highly paid culture minister with the unfortunate necessity of getting involved in grubby business. Musing for a motto for his desk, one mischievous colleague suggested 'the buck doesn't stop here'. At the end of his time at Sotheby's, Gowrie lunched with Henry Wyndham, trying to persuade him to come and take over his job. He painted such a rosy picture that Wyndham asked why he was leaving? 'Because I'm too intelligent,' Gowrie replied.[11]

The same period was marked by a changing of the guard at Christie's too. Guy Hannen and David Bathurst had modernised the firm to compete against Sotheby's in New York, and now they cast about for a new worldwide managing director based in London. Christie's owned Whites, its own catalogue printing firm, often leading to delicious confusion with the club of that name. The reorganisation of Whites revealed the skills of Christopher Davidge, who turned the firm from a small five-man operation into a successful international business on both sides of the Atlantic.* Davidge invariably delivered, and he impressed the directors of Christie's. Starting his working life by selling men's shirts at the weekend in Petticoat Lane, he retained a cockney shrewdness. In 1985, he was made managing director of Christie's, recalling that the firm was 'like the Conservative party, full of pomposity, arrogance, filled with people from a narrow social circle, who were not commercially aware'.[12]

James Roundell, head of the Impressionist Department, was initially pleased with the arrival of Davidge, because 'he backed us and allowed us to do things and empowered us to do our stuff'.[13] Although he respected the specialist departments, by degrees Davidge became more manipulative, ruled by fear, and revealed a Napoleonic streak. As one director told the author, 'Chris was undoubtedly efficient, but he had no range or emotional intelligence.'[14] With no life beyond Christie's and printing, Davidge did not appeal to his London colleagues, and it was only in America and Asia that he could be a big man. Given how counter-cultural he was to the organisation, it was inevitable that he would clash with his more *ancien régime* colleagues. As far as the new London chairman Charles Allsopp (who became Lord Hindlip in 1993) was concerned, he was 'a boring little man with no sense of humour and very chippy'.†

Allsopp, who had been made chairman the year after Davidge's appointment, was the polar opposite. A classic old-style Christie's recruit, he was the godson of a director who came straight from Eton and the Coldstream Guards. Attractive, energetic and slightly chaotic, Allsopp was popular in the firm and above all with the grand British clients whom he amused

* Davidge joined Whites at twenty-one, and within six years he was Christie's managing director.

† Interview with the author. It is worth pointing out that such baleful views are to some extent retrospective and reflect later events.

and whose language he spoke. With dashing figures like Patrick Lindsay in the Old Master Department and Charles Allsopp, Philip Hook discerned a certain 'eighteenth-century rakishness' about the place. In *Optical Illusions* (1993), his entertaining novel about Christie's, he lampooned his former colleagues. Hook's raffish anti-hero, Freddy Fairbanks – a cross between Lindsay and Allsopp – 'never accepted shooting invitations for Wednesday as it was difficult to decide which weekend to tack them on to'.[15]

Notwithstanding Hook's comic characterisation, Allsopp was shrewd and felt that Gowrie, his opposite number, was merely playing at it. In the end he was right – Gowrie never took a major auction, something at which Allsopp excelled. His relaxed and cheerful auctioneering manner was described as resembling that of an overenthusiastic evangelical preacher. Allsopp's greatest test came with the sale of Van Gogh's *Sunflowers* in 1987, but Christie's was already on something of a roll in London during the first half of the decade. In 1984, they sold seventy-one old master drawings from Chatsworth for £21 million, and in 1985, the Marquess of Northampton's Mantegna, *The Adoration of the Magi*, was sold for £8.1 million, a world record for a painting until it was shattered by *Sunflowers* two years later.

James Roundell took over the Christie's Impressionist Department at a low period in 1986. He estimated that at that time Sotheby's still had between 65 and 70 per cent of the worldwide modern picture market. In December that year, he took a group of impressionist paintings to Japan, where his brother had good contacts in the insurance industry. At dinner in Tokyo, one senior employee announced, 'My boss wants to buy pictures for the company museum', and to everyone's surprise, Roundell was asked to meet the God-like chairman of Yasuda Fire and Marine Insurance. He showed him the catalogue of items, which included a Manet and a pair of Renoirs. The chairman bought the two Renoirs and was looking for a centrepiece for the museum when Van Gogh's *Sunflowers* came up for sale the following year. This version of the subject came from Alfred Chester Beatty's collection. *Sunflowers* had a special resonance in Japan, as another version had been destroyed by American bombing in the war.

To secure the sale, Charles Allsopp had masterminded negotiations with Chester Beatty's heirs, the family of the Earl of Warwick, and he took the auction. The painting was sold on 30 March 1987, before a black-tie audience of 1,500, who filled four rooms. Allsopp opened the bidding at £5 million and, given the number of bidders, raised the bids in half-million-pound increments. At £20 million, there was applause in the room and the bidding turned into a telephone duel between two Christie's Impressionist Department directors, James Roundell and John Lumley. Roundell won the battle for his Japanese client at £24.75 million.

At nearly three times the previous world record set two years earlier at Christie's with the Mantegna, *Sunflowers* represented a step change in the art market. It was the beginning of the huge rise in auction prices that took place mostly in New York over the next decade. In retrospect, John Lumley sees *Sunflowers* as the end of London's hegemony, with the future of its modern market lying in specialist sales categorised by movements or nationality. Only rarely were major modern items sold in London after this. *Sunflowers* also represents the end of Sotheby's domination of this part of the market. Perhaps the main gainer was Charles Allsopp, who emerged as the grandee of the London art market – the 'go-to man' for Lord Rothschild.

With a symmetry so characteristic of the two main auction houses, the week after Christie's *Sunflowers* triumph, Sotheby's had its own triumph with the sale of the Duchess of Windsor's jewels. Held in Geneva, it had the fingerprints of London all over its catalogue, marketing and subject matter. The sale made nearly $50 million against a presale estimate of $7 million, making it one of the most successful sales of all time in terms of exceeding estimate. The seemingly inexorable rise in prices continued despite 'Black Monday' in October 1987, the most catastrophic fall in financial markets since 1929. The timing was potentially disastrous for Sotheby's, about to offer Van Gogh's *Irises* in New York, in a sale that drew inevitable comparisons with *Sunflowers*.

In the event, *Irises* more than pipped *Sunflowers*, selling for $54 million to Alan Bond, the Australian entrepreneur. There were sighs of relief across the art market and especially at Sotheby's, until Geraldine Norman broke the story that the firm had lent Bond 50 per cent of the purchase price. The revelation caused a storm, with cries of market manipulation. Dede Brooks, a former banker now in charge of the New York office, saw no problem with this kind of deal, but David Nash, the head of the Impressionist Department, 'recommended that we abandon lending on bids. Not because it didn't work. It worked beautifully. But because it was getting so much bad publicity.'[16] Things worsened when it was revealed that Bond couldn't keep up the payments, and a private deal was negotiated with the Getty, where the painting is today. Sotheby's was forced to stop lending money to potential buyers, although offering extended credit to favoured clients continued at both houses.

What was just gearing up was the ever-increasing habit of giving guarantees to vendors. Guarantees, introduced by Peter Wilson as long ago as the Poussin sale in 1956, were nothing new. Jo Floyd, the chairman of Christie's in 1984, had outlined his determination not to dabble in financial services which might 'provide an undue influence on demand and create an artificial price level for works of art'.[17] This was to change under Davidge, who realised that to compete for estate sales in New York, executors now regularly required guarantees as a condition of consigning. As he put it, 'We have in certain cases found it difficult to compete without offering guarantees.'[18] Thomas Gibson believed that the turning point at Christie's was when the financier Joe Lewis took a stake in the company.* Guarantees made it much harder for dealers to compete with auction houses when their only advantage was to put cash on the table, and this development increasingly shaped the top of the market.

Despite the stock market catastrophe of 1987, the years immediately after proved to be the best in Sotheby's and Christie's history. The art market continued its upward spiral until the autumn of 1990, with the rise and fall attributed to Japanese collecting. A Japanese property boom in the second half of the 1980s fed the voracious buying of impressionist and modern paintings. According to Philip Hook, by 1988, '53% of all auction sales of impressionist and modern paintings were being made to Japan. The Japanese are great consumers of brands, and "Monet" or "Renoir" were brand names that had a particular financial and cultural resonance.'[19]

Michel Strauss, head of Sotheby's London Impressionist Department, was scathing about Japanese buyers, whose taste 'ran to pretty Renoirs, of which there were truckloads, boring Monets, colourful, repetitive Chagalls, tanker-loads of paintings by the artists of the School of Paris, Utrillo, Vlaminck and, of course, their own artist, Foujita who spent almost all his working life in Paris'.[20] According to Strauss, during the five-year period from 1985 to 1990, prices in the impressionist and modern field doubled or even tripled. The market was utterly dominated by Picasso and Van Gogh, but of the top ten world records set by the end of 1989, only two, *Sunflowers* and Picasso's

* John Lumley told the author: 'The first Christie's guarantee of which I am aware was on four pictures from the estate of Robert Lehman sold in NYC, 15 May 1990.'

Acrobat and Harlequin, were sold in London – the market had effectively moved to New York.

The climax of the impressionist bubble came in May 1990, in New York, when Ryoei Saito, a Japanese collector, paid two world record prices: first at Christie's for Van Gogh's *Portrait of Dr Gachet*, which he bought for $82.5 million, and second at Sotheby's for Renoir's *Au Moulin de la Galette*, which he bought for $77 million the same week. However, by the autumn of that year, the Japanese stock market had fallen dramatically following Saddam Hussein's invasion of Kuwait. This was sudden death: the turnover of both auction houses collapsed by 60 per cent, with the inevitable staff losses. The dealers were just as badly affected.

Philip Hook recalls that 'as a London art dealer I noticed the change immediately after the summer holidays. No matter how many telephone calls you made, how many pictures you offered to your clients, at suicidally reduced prices, there was simply no business to be done anymore … days at work suddenly became very empty. My colleague Henry Wyndham and I were reduced to going to the cinema in the afternoons. That winter we saw every matinee in London.'[21] Julian Barran describes the drop in the market vividly: 'the December impressionist sale at Sotheby's in 1989 made £110 million, whereas the same sale a year later made only £10 million'.[22] He saw the loss of the Japanese as the biggest reason, but 'one must also admit that London was slowly beginning to lose its grip and there was a general recession. It was the watershed of the rise and rise of contemporary art.'[23]

It would take five years for the auction houses to return to the 1990 levels of turnover. Increasingly dissatisfied with Ainslie's stewardship, Alfred Taubman appointed Dede Brooks as CEO of Sotheby's worldwide in 1993. Brooks was the classic case of a breath of fresh air that turned stale. With a banker's brain and excellent communication skills, she was initially a cheerful, sunny presence and a champion of the staff, but her vaulting ambition and autocratic side emerged later. Taubman, who could sometimes be overbearing himself, was slightly frightened by her, although he was reassured by her prioritisation of 'shareholder comfort'. Where this would lead, we will see in another chapter. In London, Lord Gowrie resigned, and his job was divided between the Swiss *wunderkind*, Simon de Pury, for Europe and Henry Wyndham, a former Christie's paintings specialist and dealer, for Britain. They were as different as could be imagined, but very effective in their spheres. A rather tortured, driven soul, de Pury came alive as a rock star in the rostrum, shouting out bids in many languages. By contrast, the much-loved Wyndham is the quintessence of laid-back English charm and wit.

Although Sotheby's and Christie's had distinct brands, one American, the other British, it was becoming increasingly difficult to tell them apart in terms of strengths and weaknesses as they each began to occupy the other's traditional high ground. The loss of Sotheby's impressionist hegemony was dramatically brought home by Christie's winning the Koerfer sale (sold over four auctions held in New York between 1994 and 1998), which made a total of £160 million and was the biggest single consignment to any auction house to date. The Koerfers were a family with whom Sotheby's had a long relationship but, on this occasion, Christie's had tried harder, been better listeners and were more imaginative in securing the sale. By the same token, Sotheby's were stealing Christie's old master ground. A series of sales from the British Rail Pension Fund, the Dutch paintings from the Enrico Fattorini collection, the Bentinck-Thyssen collection, and the great Gentileschi *Finding of Moses* from Castle Howard gave it the edge. Sotheby's overtook Christie's in old masters in the mid-1990s. It was partly the result of an effective team led by Alexander Bell, ably backed up by George Gordon, and partly the result of the same relentless pressure to win over the British market that Christie's were applying to the modern market. Despite this, Julian Barran was right: the market that would sneak up in the night and overtake all others was contemporary, and this belonged to the dealers.

Chapter 21

The Getty Factor: Old Master Drawings

'The worldwide drawings scene was quite quiet in the 1960s … nobody could have imagined how many great drawings would become available over the next decades.'[1]

Noël Annesley

In the case of drawings, 'The market went from a quiet pursuit on the part of mostly European academics and intellectuals to an international passion embraced by wealthy collectors and a new generation of museum curators in America,' wrote Jennifer Tonkovich.[2] Out of this small, cosy world emerged one of the fastest-rising fields in the art trade, driven largely by American museums' desire to create print rooms to compete with their European counterparts. One museum, the Getty, was in more of a hurry than the others. When George Goldner became the Getty drawings curator, he was fortunate to coincide with the astonishing sale of drawings from Chatsworth held by Christie's in 1984. This was not only the greatest auction sale of drawings in memory, but also the high point of the Getty's aggressive acquisition policy that caused such a headache for the British heritage lobby.

To Noël Annesley, head of Christie's drawings and prints department and Group Deputy Chairman:

> In the 1960s it is probably fair to say that the drawings market was largely centred in London. Drawings were plentiful and still comparatively cheap. There were regular auction sales, for the most part at Sotheby's, and a coterie of knowledgeable dealers, ranging from long established specialist firms like Colnaghi's, to individuals of more recent vintage, their ranks swollen by many fine experts, often with a museum background, who had been forced to flee the horrible convulsions of continental Europe in the 1930s.[3]

To Richard Day, Annesley's opposite number at Sotheby's and later an independent dealer, the main collectors during the 1960s – in his unflattering characterisation – were 'the connoisseur Count Antoine Seilern, the irascible badger Colonel Colville, the devious Robert Lehman and the unlovely Baron Hatvany'.[4]

During the 1950s, Jim Byam Shaw and the émigré dealer Hans Calmann emerged as the two key players in the transformation of the London market, the former from the venerable firm Colnaghi and the latter starting from scratch. Byam Shaw was the most widely respected specialist in Bond Street, and the epitome of the gentleman scholar dealer. In Noël Annesley's words, he 'set a standard for all of us who, in one capacity or another, were engaged in researching drawings, cataloguing them, and no doubt most satisfying of all, collecting them'.[5] Asked the difference between the two dealers, Annesley quipped that, 'if you had good taste you went to Byam Shaw, and if you were more adventurous or more optimistic you went to Calmann'.[6]

Calmann was a stockbroker and collector in Hamburg who came to London in 1937 and was interned on the Isle of Man during the war. He set up as an old master drawings dealer at 15 Davies Street, relocating to 39 Bruton Place in 1963. Describing the London scene, Calmann claimed, 'There was only one big

London firm dealing in drawings, Colnaghi's … otherwise practically nobody else save one or two refugees who found the cheapness of drawings and the chances they gave to an inquisitive mind most suitable for their slender means.'[7]

Drawings dealers, like those in old master paintings, tried to add value through knowledge. Calmann's initial strategy was to deal in inexpensive drawings he knew would be easier to sell. In his unpublished memoirs, he records his increasing income: £6,000 in 1945 up to £18,000 in 1960. Early on, he expected to make only £10 or £20 on each sale, but by the 1960s he got into his stride and charged two or three times the amount he paid for each drawing. Calmann made many discoveries: in 1953 he saw a 'nondescript' group at Wildenstein's London gallery, which included one superb red chalk drawing. Buying them for £1,300, he took the exceptional one to A. E. Popham at the British Museum, who confirmed his suspicion that it was a Raphael. The museum purchased it for £2,000, which, as Calmann put it, 'was not only a good profit, but beyond that an outsized feather in my cap, to have bought something from Wildenstein's and made a profit'.[8]

Calmann's name appears frequently in the provenance of drawings at American institutions such as the Morgan Library, where he placed 106 drawings. He could be imperious, and some American curators felt more comfortable dealing with Byam Shaw. Calmann's posthumous reputation was tarnished by his association with Eric Hebborn, the artist, dealer and talented forger who worked from Rome until he was murdered in 1996, not long after his activities as a forger were unmasked. Fakes 'are a real hazard in our field,' commented Annesley, and Hebborn caught out many in the drawings world, including, on one occasion, Jim Byam Shaw. For his part, Hebborn admired Calmann, claiming that 'I have heard Hans described as "overbearingly autocratic" and maybe he was, but he had an eye for quality and in spite of – or perhaps because of – his arrogant faith in his own judgement, some of the very best old masters on the market passed through his hands.'[9]

Drawings dealers did not expect collectors to challenge their supremacy in the saleroom and tried to make them bid through the trade. In fact, most collectors found it was expedient to do so, as they could probably buy items cheaper by leaving the bids with their favoured dealer. When George Abrams, the American collector, announced he would be bidding himself, Calmann warned him, 'Gentlemen do not bid at auction and I doubt if you will be successful.' Abrams opened the bidding with a smile at Calmann but soon dropped out, leaving the bid with the dealer, who smiled back. At that point, Herbert Bier, another distinguished dealer, came into the bidding and made the winning bid, which, it turned out, was on behalf of Abrams, who never missed a sale after that.[10]

The drawings world had the most rigorous training in the art trade. Before joining Colnaghi, Jim Byam Shaw and John Baskett spent many years studying in European print cabinets. When Richard Day was offered the job as drawings cataloguer at Sotheby's, as part of the conditions he was allowed to spend time every week at the British Museum's Department of Prints and Drawings and was given two months abroad every year to study in the major European public collections. The salary, however, was only £8 a week. Day was apprenticed to a lovable Pickwickian figure called Mr F. L. Wilder, whose main interest was in sporting prints.*

Mr Wilder had joined Sotheby's in 1911 and still came in most days in the 1980s. He had watched the rise and fall of the mezzotint market in the 1920s and 1930s and the collapse of the etching revival around artists such as Muirhead Bone and F. L.

* Mr Wilder came to work on the bus every morning from Woodford, which took over an hour, so he read detective stories on the way. In 1993, he celebrated his 100th birthday with a party at Sotheby's.

Griggs. He had also watched the enormous increase in value of Rembrandt and Dürer prints, while so many other printmakers were overlooked.[11] The real growth market was for nineteenth-century and modern prints: a notable success was the Ludwig Charrell collection of Toulouse-Lautrec lithographs. Sold at Sotheby's in 1966, it was the first time that prints had been given a hard-back catalogue.

Annesley describes how, when he joined Brian Sewell at Christie's in 1964, they 'sought to challenge the widely held assumption that if you needed to sell drawings or prints you went to Sotheby's'.[12] The auction catalogue entries were minimalist and often did not commit to an artist or even a date of execution. There were no printed estimates, no glossary or explanation of cataloguing terms, and condition reports were unheard of. As Annesley put it, 'a participant was expected to know how it all worked', and thus it was a world of professional dealers. Annesley was fortunate that David Carritt had just discovered a volume of Domenico Tiepolo drawings in Earl Beauchamp's collection, which Christie's sold in 1965. That and the sale of Italian drawings from Harewood House the same year persuaded Christie's management to give the department a separate identity which put it on equal terms with Sotheby's. Sotheby's had a coup the following year when they hired Philip Pouncey, the greatest drawings connoisseur of his generation. As Annesley admitted, 'it was a great feather in their cap although he was only interested in Italian drawings'.

Later the doyen of London drawings specialists, Annesley was targeted early on by Hebborn, who relied on the young man's inexperience and eagerness to fill sales against heavy competition. In 1968, he fobbed a 'Pontormo' onto Annesley, which has been the property of Christie's ever since. Sotheby's probably held the upper hand during the 1970s with sales of Ellesmere (1972), Gathorne-Hardy (1976) and Robert von Hirsch (1978). Seven drawings in the von Hirsch sale making over a hundred thousand pounds each was the first indication that something was stirring in the drawings market. The sensation was a Dürer watercolour landscape which made £704,000, almost quadrupling the previous record for a drawing, which would only be broken at the Chatsworth sale in 1984. If the 1970s was Sotheby's decade, the 1980s was the decade of Christie's triumphs, starting with the Hatvany sale (1980). It was, however, the Chatsworth drawings sale that would capture the attention of the world.

In August 1982, along with Christie's tax advisor Christopher Ponter, Annesley was summoned to a meeting with the lawyer of the trustees of the Chatsworth Settlement. The trustees wished to create a £5 million endowment fund, raising this sum from the celebrated old master drawings collection. Consisting of over 2,000 sheets, it had been formed in the eighteenth century by William Cavendish, 2nd Duke of Devonshire. The trustees, influenced by Andrew, the sitting Duke of Devonshire, recommended a balanced selection to appeal to an institution who might wish to buy the drawings as a group. Northern museums, and in particular the Whitworth in Manchester where the duke held an honorary position, were favoured. In anticipation of the potential negative publicity and heritage fallout, the trustees preferred a sale to a UK institution. Aside from the Inigo Jones drawings, which were off-limits, Annesley aimed to select each drawing so that a comparable example remained in the collection.

Drawing up a list of seventy drawings, Annesley went to Chatsworth to check their condition. His valuation came to just short of £6 million. In October 1982, he delivered a proposal to the Arts Minister and heard nothing until the following April. The answer came back that the British Museum, rather than a northern museum, must be the recipient. A meeting took place in June with all the interested parties. In the meantime, Annesley had revised his valuation to £7.2 million, owing to a recent surge in prices. The only body that could fund such a purchase was the newly formed National Heritage Memorial Fund. The chairman of the fund, Lord Charteris, consulted his friend, Geoffrey Agnew – whose firm specialised in paintings – on the valuation of the drawings.

Sir David Wilson, an archaeologist and director of the British Museum, was never enthusiastic, and John Rowlands, the keeper of prints and drawings, and his advisors at Agnew's queried some of the attributions and values, suggesting a figure £1.75 million lower. For instance, Christie's valued the 'Vasari' sheet (consisting of ten drawings by early Florentine artists mounted by Vasari) for £750,000, and Agnew's for only £250,000. In the sale it was to make over £3 million. Matters dragged on and the British Museum came up with a final offer of £5.25 million. Although the difference was not enormous, the offer was unacceptable to the trustees. As it turned out, the gap was, as Annesley put it, 'very good for Chatsworth, Christie's, and my career'.[13] Four months before the sale, Annesley and Ponter drove to Chatsworth to collect the drawings, and in those relaxed days, put the boxes in the boot of their car and set off back to London in a snowstorm.

The sale that followed would be dominated by one man: George Goldner, the Getty drawings curator. To observe the Getty effect on the London art market, the Chatsworth drawings sale provides the seminal example. With Paul Getty's death in 1977, the greatest surprise was that his holdings in Getty Oil were left to his previously starved museum in Malibu. Its transition from one of the poorest museums in America to by far the richest was challenging. The founder left no instructions or vision for how it should develop. Even though the museum had unimagined wealth, the lack of clear direction, and the many conflicting views at different levels of the administration, initially made for a halting and often confused strategy. Parts of the collection that Getty bequeathed, notably the French furniture and antiquities, were very fine, but the paintings collection was disappointing.

To Burton B. Fredericksen, the museum's birth 'could be compared to that of a child conceived out of wedlock, an offspring about whose origins the parent felt some uneasiness'.[14] Like the owner's children, the museum 'was left to be brought up by others and so would come to lead a stunted and unstable existence for much of its early life'.[15] By the 1980s, unprecedented funds were available and the decade was to be a golden one for acquisitions, many from British collections, which raised anxieties in the heritage and museum world. In May 1984, Lord Charteris warned the House of Lords about the threat the Getty posed to the nation's artistic heritage. By that year, the endowment had risen to $2.3 billion and the museum was in a position to buy almost anything it wanted. It was at this propitious moment that the Chatsworth drawings appeared.

An art historian and collector of drawings, George Goldner joined the Getty in 1979 to develop the photographic archive. He made the case to his trustees that it was still possible to make a collection of drawings while in other areas it was already too difficult to find material. Determined, persuasive, energetic, intellectual and passionate, Goldner founded the drawings department in 1983, the year before the Chatsworth sale. His acquisitive urge and achievement were unmatched in modern times and he came to utterly dominate the market for old master drawings. As Annesley recalled, 'the Getty drawings collection was formed from scratch'. Goldner began by buying Rembrandt's *Cleopatra* as early as 1981, and paused, so that wags used to describe the collection as 'the Getty old master drawing collection'.

Before the Chatsworth sale, Goldner went to see David Wilson and his keepers at the British Museum and made an astonishing offer. He suggested that the Getty purchase the entire collection at the asking price of the Chatsworth trustees, and then split the drawings 50/50 with the museum at no cost to themselves. Goldner's only condition was that there would be no export problems with their half. The offer was rejected.*

The excitement generated by the Chatsworth drawings sale was intense and attracted major collectors from outside the field. Goldner chose a small group of dealers to act on commission for the Getty: Richard Day, Herman Shickman, Adrian Eeles, the Munich dealer Bruce Livie of Arnold-Livie and John Morton Morris of Hazlitt's. Richard Day had a grey flannel suit specially made for the occasion and booked himself a suite at Brown's Hotel. He arranged lunch in a booth at Wilton's to meet George Goldner and discuss tactics. In the next booth, Adrian Ward-Jackson was entertaining Basia Johnson for exactly the same reason.[16]

Of course, the problem arose that most of these dealers were also asked to bid on the same lots by other museum directors, like Charles Ryskamp at the Morgan. Collectors like Ian Woodner, a New York property developer who formed an important group of drawings, most of which he gave to the National Gallery in Washington, were also eyeing up the opportunities. Goldner's main competition would turn out to be private collectors like Woodner and Basia Johnson.

The sale was held in the evening on a very hot day in early July and the lights from the TV cameras made the heat uncomfortable. Goldner placed his various dealers around the Great Room at Christie's. Richard Day, seated on the left aisle about halfway down from the rostrum, was disconcerted to find the important New York dealer Eugene Thaw in the seat behind him. Annesley felt nervous as he climbed into the rostrum before a

* Information from George Goldner. However advantageous to the British Museum, the offer would have circumnavigated the rules of the Export Reviewing Committee.

packed house, with two ancillary salerooms equally full. The first major lot, Federico Barocci's *Entombment*, was bought by Bruce Livie on behalf of the Getty for £388,800, five times the estimate. Annesley relaxed; it 'was going to work,' he thought, as the drawings doubled and tripled their estimates. Day bought the Mantegna for £1,188,000 for Goldner, but the sensation of the sale was Raphael's *Study of a Man's Head*, bought by Basia Johnson for just over £3,564,000. By all accounts, this upset Goldner very much, and he also had stiff competition from Woodner on other lots, including the Vasari sheet. Despite this, he was able to buy the other Raphael in the sale through John Morton Morris for £1,512,000, and the Holbein portrait drawing for £1,595,000.

Goldner didn't have enough money to buy all the things he wanted, but he was able to buy seven of the best drawings in the sale, including works by Rubens, Van Dyck, Mantegna, Holbein, Rembrandt and Raphael for just over £6 million, an extraordinary figure for the time.* This was not the end of the story, however, as Goldner had persuaded dealers to buy the drawings he could not afford on the day on the promise that he would buy them as soon as he could raise the money. Over the following two years, he bought a further thirteen drawings from the sale, making a total of nineteen additions to the collection.

Richard Day observed that, 'George's orchestration of the sale was masterly. He had faced challenging times to persuade the Getty trustees that his proposed bids (far above the Christie's estimates) were realistic, but his figures turned out to be accurate. With the museum's purchases … the Getty became overnight a significant place to see drawings of a quality, though not a quantity, to rival the principal drawing cabinets of Europe.'[17]

The seventy-one lots from Chatsworth made £21 million, four times the British Museum's last offer. The sale transformed

* The Rembrandt was stopped at export and is now in the British Museum.

the drawings market and afterwards Sotheby's and Christie's extended drawing sales from London and New York to Amsterdam, Monte Carlo and later Paris. The British heritage establishment was in turmoil at having made such a miscalculation before the sale. Geraldine Norman ran a piece in *The Times* called 'One more bungle denting Britain's heritage'.[18] Most of the Getty purchases were given a temporary export stop. The dealers concerned had to go and make the case to the Export Reviewing Committee, which Richard Day, with an element of hyperbole, described as 'My idea of the British equivalent of the KGB headquarters, the Lubyanka. You faced a formidable row of expert art historians and have a large secretariat taking notes while you made your case for export.'[19]

In the annual report on the acquisitions of 1984, John Walsh, the Getty director, went to some length to try to dispel the notion that 'Getty takes all' by emphasising that they had only bought six out of seventy-one drawings in the sale. He also underlined the museum's intention 'to buy selectively, patiently, and with attention to our potential effect on the market'.[20] Goldner's activities manifestly demonstrated the inaccuracy of this statement. The Getty buying spree continued across the board, and Burton Fredericksen stated that 'greater sums were expended by the Getty on the purchase of paintings during 1985 than by any other institution or collector in modern times over a comparable period'.[21] Of the thirty paintings acquired that year, the lion of the group was Mantegna's *Adoration of the Magi*, bought at Christie's for £8.1 million. No British institution could match the sum, and the export licence was granted. At the end of the decade, the Getty carried off Adriaen de Vries's bronze *Dancing Faun* (see Chapter 13), but failed to secure an export licence for Canova's *Three Graces* when the V&A and National Galleries of Scotland pooled their resources, aided by a donation from London-based Paul Getty Jr. In the following decade, the National Heritage Lottery Fund, set up in

1994, boosted the ability of British museums and galleries to secure items the Getty tried to export. These included Raphael's *Madonna of the Pinks*, saved for the National Gallery after a public appeal in London in 2004.

The drawings market continued to rise and rise except in one area, as changing tastes led to the disappearance of the British watercolour market, with the major exception of Turner. Once the preserve of so many enthusiasts and the subject of winter exhibitions at Agnew's, Leger and other dealers, for decades British watercolours were a major preoccupation of British collectors and Bond Street dealers. Out of the ashes of this market, kept alive by Andrew Wyld and Lowell Libson, Turner became ever more sought-after by international collectors. On the old master front, London nurtured a group of talented drawings specialists, notably Jean-Luc Baroni, Richard Day's partner James Faber and one of the most interesting drawings dealers, Katrin Bellinger, a sometime trustee of the London National Gallery. Her role models in this field were her predecessors Kate de Rothschild and Yvonne Tan Bunzl.

Where the best items in other markets were sent to New York for sale, drawings remained strong in London. This was well demonstrated at the turn of the millennium when both Christie's and Sotheby's salerooms offered a Michelangelo drawing in London. In 2000, Christie's sold Brinsley Ford's *Study for the Risen Christ*, for a world record of £8,134,750. The following year, Sotheby's sold *A Mourning Woman*, discovered by Julien Stock in the library at Castle Howard. Bought by Baroni, the drawing fetched almost £6 million.

Chapter 22

Antiquities: The Gathering Storm

'It was the MeToo moment for art.' *

Rena Neville

No market was more of a rollercoaster than antiquities. It was not just the price fluctuations, but the changing ethical climate and scandals that engulfed it. Antiquities were undervalued in the 1940s and 1950s, and the transmission of ancient culture was largely in the hands of archaeologists.† Early twentieth-century avant-garde artists, and art historians like Roger Fry, had made known their preference for the power of tribal art. However, the cultural relativism which saw tribal art and classical objects as part of the universal aestheticism of André Malraux, amongst others, led to a reappraisal of antiquities by the end of the century. Hitherto appreciated either in the sculpture gallery or in context at the archaeological dig, they increasingly began to be admired as totemic works of art to be displayed alongside contemporary art. The dealer John Hewett conveyed this taste to Robert and

* Rena Neville, first head of compliance at Sotheby's.

† Kenneth Clark in *The Nude* (1956), p. xxi, lamented, 'The dwindling appreciation of antique art during the last 50 years … and professional writers on classical archaeology, microscopically re-examining the scanty evidence, have not helped us to understand why it was that for 400 years artists and amateurs shed tears of admiration for works which arouse no tremor of emotion in us.'

Lisa Sainsbury, and it is seen today at the Sainsbury Centre, Norwich, which displays small-scale examples of ancient art alongside the works of Francis Bacon.

From a rather quiet post-war market, occasionally enlivened by the sale of a country house Grand Tour collection, antiquities soared in value during the 1980s, under the high-powered salesmanship of Hewett's protégé, Robin Symes. American museums and collectors entered the market, creating a demand that in turn stimulated the '*tombaroli*', or tomb raiders, who formed part of a complicated distribution web involving several international dealers and entangling many collectors, museums and auction houses, including Sotheby's. This illegal trade, with the concomitant dislocation of objects from their context, caused archaeologists to raise what became increasingly widespread concerns around collectors and art dealers who encouraged the illegal looting of sites by handling works without provenance.

Before these revelations, it was an untroubled market into which J. Paul Getty stepped in 1951, when he came shopping in London. Getty realised that, thanks to post-war social changes in Britain, there were extraordinary opportunities to acquire items from country houses. By far the most important item he acquired on this visit was the Lansdowne Heracles, a full-length marble statue excavated near Hadrian's Villa at Tivoli and sold to the Marquess of Lansdowne in 1792. The statue was on sale at Spink's for £10,000, and Getty successfully offered £6,000. The purchase would have enormous significance for the direction of his collecting and the villa museum he had founded in Malibu.*

Spink was the biggest dealer in antiquities until it was eclipsed by Robin Symes. An undated Spink catalogue from

* J. Paul Getty wrote a novella, *A Journey from Corinth* (1955), in which there appears a statue that can be identified as the Lansdowne Heracles.

the 1930s shows the firm's range – modern jewellery, sixteenth-century Italian bronzes, paintings by Reynolds and Morland, Japanese prints – but the most expensive item was an antiquity, 'the famous bronze statue from Lake Nemi', priced at £25,000. Spink had interesting British clients, from the actor Ralph Richardson to the critic Denys Sutton, while Christopher Cockerell, inventor of the hovercraft, liked buying Greek vases. Although Spink advertised every week in *Country Life*, more adventurous collectors and dealers were drawn to John Hewett.

During the 1960s, Hewett was the low-key king of the London antiquities market. He may not have had the big pieces that Spink handled, but he was far more influential. Having almost a cult following among collectors and young dealers starting out in the trade, he was described as a shaman-dealer.* Hewett trained Bruce Chatwin, and was the main influence on Robin Symes, the future leader of the trade. Discussed in a previous chapter as the central figure in the tribal art story, Hewett dominates the early part of this chapter. It was during the war when he was posted to Naples with the Scots Guards (rising to be a sergeant) that he discovered the ancient world and bought his first pieces. His shop in Sydney Street, off the King's Road, displayed an eclectic mix of ancient objects and African masks. Although mesmeric to some, other visitors had the impression that he was reclusive, and one described him as 'a short, steady, lightly bearded man in a tweed suit, smoking a pipe and carrying a walking stick'.[1] Combining antiquities and tribal objects with contemporary art, Hewett sold Giacometti sculptures to the Sainsburys, and Bacon did a triptych of him.

The American academic turned dealer Bob Hecht recalled visiting Istanbul with Hewett in January 1962, where they

* Hewett was the dealer most admired by Anthony d'Offay.

bought a Roman bronze head.* Hewett carried the head in his hand luggage through the airport and sold it to the Geneva collector Martin Bodmer. His main Geneva client, however, was George Ortiz, who formed an extraordinary collection of tribal art and antiquities with little regard for provenance. Collections such as the Sainsburys' and Ortiz are Hewett's most visible legacy, but much of his influence came from setting up the Sotheby's Antiquities Department for his friend, Peter Wilson.

Hewett ran the Sotheby's department at arm's length, acting as consultant. In contrast to the Christie's department, from its earliest days the Sotheby's department depended largely on dealer consignments, which would have long-term consequences. Bruce Chatwin, the future novelist, was Hewett's first cataloguer. Seeing them together, one dealer observed, was 'like watching a young puppy with a silent block of marble'.[2] Chatwin would not stay long at Sotheby's, and on his departure, Felicity Nicholson, his former secretary, described as 'Irish, clever, intuitive and a little dotty',[3] would take control of the department.

Whilst at Sotheby's, Chatwin entered an informal dealing partnership with Robert Erskine, an Old Etonian coin and antiquities specialist. A dashing figure who ran the St George's Gallery, Erskine was regarded as scholarly and honest. He provided the money and Chatwin the Sotheby's client list, and they shared the profit on the objects they sold. Like most antiquities dealers who would spend the winter going to various countries to find material, Erskine and Chatwin travelled to Egypt together to collect items for sale. Cultural laws did not yet exist in many places, and it was still possible to export most things

* Hecht was a mix of scholar, connoisseur and adventurer, who held views that are now highly contested. He believed that if the objects were not exported, they would be hidden from wider public view in local collections.

from Egypt up until 1983. Chatwin was, however, ordered to break up his arrangement by Sotheby's since it constituted a conflict of interest. Erskine went on to make television programmes on antiquities, and covered the two worlds of archaeology and art dealing before they became antipathetic.

In 1964–5 Sotheby's and Christie's held two landmark auctions in London that were characteristic of their differing client base. In 1964 Sotheby's sold the collection of the former New York dealer Ernest Brummer, which consisted of Egyptian and near Eastern antiquities, in an early example of a one-owner catalogue. In 1965 Christie's sold objects belonging to Captain George Spencer-Churchill from Northwick Park, part of a much wider collection of old master paintings and other material. Although Christie's hoped it would be the first £100,000 antiquities sale, it made a very respectable £96,000. These sales set the tone for the next twenty years, during which Christie's dealt with British aristocratic collections and spurned dealers' property. This was partly because they had the regular supply of country house material, notably antiquities from Wilton (1961), a major Grand Tour collection.* It was also because Christine Insley-Green, their specialist from 1979, was an archaeologist by training and very alert to looting. 'We didn't want to get too pally with dealers,' as Judith Nugée told the author, in particular Robin Symes, who, according to Richard Falkiner, was regarded at Christie's 'as a jumped-up parvenue'.[4] At Sotheby's, by contrast, Felicity Nicholson became very close to Symes, who found her *molto simpatico*, and they often dined together.

The price rises for antiquities from the 1970s to the 1990s are evident across a series of auction sales. The first indication of a great change in the market came with the Constable-Maxwell collection of ancient glass, held at Sotheby's in June 1979. Lot 41, the Roman glass Cage Cup, made £520,000 and was bought

* This sale was organised by their specialist at the time, Richard Falkiner.

by the British Rail Pension Fund (which sold it in 1997 for £2.1 million). This was followed the next year by the sale of eighty vases from the collection of the Marquess of Northampton at Christie's. However, it was Sotheby's record-breaking sale of the Comtesse de Béhague's vast collection of Egyptian, Greek and Roman antiquities, alongside Byzantine and early Christian pieces, held in Monaco in December 1987, that ignited the market.

The antiquities dealer James Ede described the Béhague sale as 'the greatest sale in my lifetime'. The first black-tie sale for antiquities, it was a glamorous event, arranged and catalogued from London. Oliver Forge recalls that, 'It was the time when Robin Symes was moving into a higher level of dealing and really began to dominate the market on both sides of the Atlantic (John Hewett had retired by then). The richest American collectors were at their most acquisitive and that was when antiquities started to generate a lot of interest and make serious money.'[5] The trend towards price inflation was underlined by the spectacular price achieved for the Assyrian relief from Canford School, which made £7.7 million when it was sold by Christie's in 1994. Within four years, Symes was asking three times this price for a statue of Aphrodite. Behind this apparent success, trouble was brewing.

In 1985, Peter Watson, a saleroom correspondent on *The Observer*, had been invited by Brian Cook, keeper of the British Museum's department of Greek and Roman Antiquities, to come and see him about something 'pretty hot'. Cook handed Watson a Sotheby's catalogue and said, 'There, that's your story. Sotheby's is selling a whole batch of smuggled antiquities.'[6] The upcoming sale, scheduled for 9 December 1985, contained among other things a black-figure Attic amphora, which the British Museum had wanted, but had been put off by lack of provenance. The meeting lit a fuse which would continue to burn over the next decade as a new climate of archaeological disapproval emerged.*

Watson focused his attention on Sotheby's, and evidence came from an unexpected source: James Hodges, an administrator in the antiquities department. Hodges was accused by the auction house of stealing two valuable antiquities, forging papers which allowed the theft, and eight instances of false accounting. He had taken the precaution of photocopying three suitcases worth of Sotheby's internal documents revealing the department's unethical behaviour as his insurance policy. Refusing to make a deal, in 1991 Tim Llewellyn, Sotheby's managing director, insisted on Hodges going to trial.

Hodges contacted Peter Watson, to whom he handed the hoard of incriminating documents. His motives were to tell his side of the story before the trial, hoping that the resulting publicity and scandal would influence the verdict in his favour and deflect the spotlight towards Sotheby's. He also wanted a share in the proceeds of anything Watson might publish. The problem was that Hodges had stolen the documents, and Watson realised that publishing before the trial could be seen as prejudicial to the case. He decided to keep his powder dry until he could tell the full story later – on television and in a book.

To return to the initial accusation of Hodges stealing two antique objects from Sotheby's, an odd event occurred one Sunday in 1989. As the collection plate was handed round at mass at the Brompton Oratory, instead of money, Hodges placed a note which read: 'Father, please give this to detective

* The first sale impacted by the new climate was the Erlenmeyer sale of Cycladic items, a collection bought mostly from the French market between 1940 and 1960 and sold at Sotheby's in June 1990. The cover lot (lot 137), a Cycladic torso, was withdrawn and sold privately to the Dolly Goulandris Museum in Athens, the only way that Cycladic pieces could be 'washed' of their recent past.

Sergeant Quinn at West End Central Police Station'.[7] The envelope disclosed the whereabouts of the stolen antiquities, stored in a luggage locker at Marylebone Station wrapped in copies of the *Racing Post*.[8] Despite having recovered the objects, and against the advice of the in-house lawyer who saw trouble ahead, Sotheby's was keen for the prosecution to go ahead.

What the Hodges trial revealed in court was the trail that ran between the Mediterranean, Switzerland and London, where dealers would place looted objects in auction. Hodges's former boss, Felicity Nicholson, head of the Sotheby's department, admitted that smuggling was widespread, and that she turned a blind eye to it. Overlooking the issue of complicity, she claimed that she did not regard it as Sotheby's problem to look after its clients' morals or business practices.[9] Indeed, evidence in one document stolen by Hodges was particularly damning for Nicholson, revealing that she found 'the shady side of the antiquities market not uncongenial'.[10] Hodges was sentenced to nine months' imprisonment, but the full impact on the antiquities trade was not felt until he came out of prison.

The prevailing view in the trade was that an embargo on objects lacking provenance would drive it underground. Tim Llewellyn stated in court, 'We can't be responsible to police the market, though it is reasonable to ask: if we were to withdraw from the market, would it solve the problem? It would not.'[11] Some in the trade saw the problem as emanating from the draconian laws of Italy and Greece, which automatically gave the ownership of excavated items to the state. It was convenient for Sotheby's to assume that their clients had title to whatever they were selling, even when they might have had cause to doubt it.

The 1994 Royal Academy exhibition of George Ortiz's collection caused a public row between the unregulated trade of antiquities and that of archaeology. It was estimated that 23 per cent of objects in the exhibition had no provenance at all, a further 62 per cent had 'said to be' provenance, and only 15 per cent had provenance that would pass today's benchmark. Colin Renfrew, a trustee of the British Museum, criticised Ortiz for indirectly financing large-scale looting, the ultimate source of so many of the exhibits. A few days after the exhibition opened, Ortiz and Renfrew had a heated debate on BBC television, in which Renfrew drew attention to the way sensitive sites were being despoiled, with objects illicitly removed and trafficked by the art trade to collectors like Ortiz.

The 'don't ask, don't tell' ground on which much of the antiquities trade stood is evident in Brian O'Connell's claim that, 'It was universally accepted that asking an art dealer where he sourced an object was as impertinent as asking a doctor to produce the references in the medical textbooks on which he based his diagnosis of a patient.'[12] To Renfrew, however, this simply made clear the fact that, 'It is generally prudent to follow the principle that unprovenanced antiquities are likely to be looted antiquities.'[13] The ongoing complicity of many involved in the antiquities trade led Richard Elia to claim that 'collectors are the real looters',[14] as the battle around trafficking artefacts became increasingly fractious.

When Hodges came out of prison in 1992, he and Watson began their collaboration in earnest. The trial had been a taster but had not captured the public's attention. However, Watson's programmes on Channel 4's *Dispatches*, and his subsequent book, *Sotheby's: Inside Story* (1997), were to have a seismic effect on the market. To spice up his story and underline his point, Watson arranged a sting. Given the recent uproar, it would be difficult to get anybody on secret camera talking about antiquities, but nobody in the paintings world would suspect anything. In March 1996, an attractive Australian woman walked into Sotheby's Milan, and persuaded the young old master specialist to accept a picture for sale in London, which it was clear she was going to smuggle out of Italy without an export licence. Channel 4's first *Dispatches* programme carried the story.

Watson's revelations were screened in February 1997. They caused a sensation. As Robert Lacey put it, 'It was so satisfying to see a lofty institution like Sotheby's caught red handed at such an unlofty game.'*[15] Sotheby's was given an opportunity to take part in the programme but offered 'no comment'. The real meat of Watson's programmes was the exposure of complicity by Sotheby's with a number of antiquities smugglers. *Country Life* ran a leader 'Del Boy Comes to Sotheby's', and the story made headline news. To make matters worse for the firm, the scandal broke in the middle of shooting an eight-part BBC fly-on-the-wall documentary (as such things always seem to happen). The American CEO, Dede Brooks, flew in, and the press very much enjoyed commenting on her big hair and suits alongside the unfolding scandal. The immediate result was that Sotheby's ceased selling antiquities in London, effective from after the British Rail Pension Fund sale, already scheduled for later that year.

After *Dispatches*, compliance became king. Mandatory provenance helped to increase accountability and make cataloguing more stringent. Alfred Taubman, Sotheby's owner, used to complain that the scandal cost him over $10 million. The first head of compliance at Sotheby's, Rena Neville, subsequently called the shift 'The MeToo moment for art', and the legal departments of Sotheby's and Christie's ballooned. Although Watson never investigated Christie's, it is unlikely that he would have found as blatant a case of collusion as at Sotheby's. One retired employee thought that 'Christie's were doing it as well, but they were more discreet'.[16] The long-term effect of *Dispatches* was the

* The Milan affair was a press coup, although it later emerged that the picture did not need an export licence.

raising of sale thresholds in London. The legal scrutiny required added an extra cost to each lot, making this inevitable, and New York had been pressing for such a rise for some time. The added cost marked the point at which Sotheby's began to lose the middle market.

After a series of public scandals, it is easy to think that the entire antiquities market was amoral, but this was certainly not the case. What constituted good practice changed dramatically in the 1980s, but even before this point, many excellent dealers plied their trade, and were unfairly tarnished by the scandals that erupted around Sotheby's and Robin Symes. Before looting became the issue, the main problem in the market was fakes. As Burton Fredericksen noted about the Getty, 'The collection of antiquities in the spring of 1973 was notable almost as much for the pieces returned to the vendors as for those acquired.'[17]

The London training ground for antiquities dealers – as in so many other areas – was Portobello Market. Most, like Symes, soon moved to their own premises, but Julia Schottlander was a typical market trader who ran a reputable stall selling items with good provenance, often to museums, and even gave material to the British Museum. Rupert Wace started as a porter in Sotheby's Belgravia, and went to work for Christie's, before opening his own shop on a first floor in New Bond Street, next to Sotheby's. He described Bruce McAlpine and Robin Symes as 'the two big players when I started'. Wace left the top pieces to them but pointed out that 'there was lots of fun to be had in the middle range', where there were discoveries to be made.[18] One day, Wace went to a minor auction in Surrey to buy an Egyptian cat. While he was there, he saw an item catalogued as 'Man with a Hat', which he recognised as a seventh-century BC Sardinian bronze. He bought it and found that it had been published in the 1920s and was previously in the collection of Oscar Wilde. Later, Rupert Wace sold a few pieces privately from the British Rail Pension Fund. The fund got caught on the cusp between the changing attitude towards provenance, and some of the pieces they had bought (often from Symes) could no longer be accepted at auction.

James Ede inherited a dealership from his father, Charles Ede, who had made his fortune setting up the Folio Society in 1947. When Charles sold it in 1971, he used the money to set up as an antiquities dealer. He was to his fingertips a mail order operative, and liked producing catalogues, which meant he was never really a shop or gallery man, despite having a Brook Street premises. Ede took the principle of selling Folio books into selling art, offering good examples at the lowest possible price. Initially, he sold quite a lot to Commonwealth museums, supplying low-cost educational items up to a limit of about £2,000. Ede sourced material from fellow London dealers and buying trips to Europe but – concerned about provenance – he never bought in Italy, Greece or Egypt. He strongly advised his son James not to follow him into the antiquities trade, believing that 'this business is finished'. When James nonetheless joined the business in 1977, he wanted to raise their game and handled the collection formed by Gustave Mustaki, a Greek living in Alexandria, who had used his collection of about 10,000 objects as a dowry for his daughter when she married an RAF officer.

The main competitors to Symes were Bruce and Ingrid McAlpine, who had the entrée to the country house market, which Symes did not. More old school and careful, selling good-quality material, the McAlpines were equal partners in the business, until they eventually separated. Before this, they had a small gallery in Clifford Street and eventually bought a space opposite Claridge's, where they became the second stop in London after Symes.

Collectors were not put off by the *Dispatches* scandal, and a rich, determined group of buyers dominated the period up to the millennium. The most acquisitive of these was Sheikh Saud bin Mohammed al-Thani of Qatar, whom some thought messed up the market because he was so capricious in his bidding. One of his acquisitions was the Jenkins Venus, at Newby Hall in Yorkshire since 1765, making it a perfect piece of provenanced Grand Tour art, which he bought from Christie's in 2002 for £7,926,650. In the wake of the scandals, a new market had been established, much more careful in its acceptance of property.

Chapter 23

The Rise and Fall of Robin Symes

'The market in antiquities is perhaps the most corrupt and problematic aspect of the international art trade.'[1]

Marion True

Of all the dealers in this book, Robin Symes (1939–2023) had the most extraordinary trajectory. Coming from nowhere, he developed the antiquities market from a few thousand dollars up to $30 million apiece. For two decades during the 1980s and 1990s, he virtually owned the market, and nobody was better able to project antiquities to a contemporary audience. In Rainer Zietz's opinion, Symes and his partner, Christo Michaelides (1945–1999), freed antiquities from dry archaeology, 'seeing them not in a Grand Tour gallery context but as independent autonomous works of art, that turned fragments into works of creative imagination'.[2] Symes presented and sold these fragments as if they were Picassos or Henry Moores. In the process, he gained the trust and patronage of a group of rich collectors and museums in America, captivated to the point of blindness (whether naively or intentionally) by the 'finds' that emerged as the market reached its stratospheric levels.

A respectable, hugely successful and stylish dealer, Symes was, however, leading a double life as the front for the massive looting industry that underpinned the antiquities market. For him, hubris was followed by nemesis, and nobody rose higher or fell faster. In many ways, his story is less a Greek tragedy than a Jacobean revenge play, as Christo's family set in motion the events that led to his destruction. The case against him ended with a squalid trial where Symes showed himself a stranger to the truth and became almost deranged in his perjury. Most accounts of his career dwell on his downfall, but equally interesting is his rise to a position of international preeminence in a field in which everybody trusted his opinion. Today, a Symes provenance would set off alarm bells, and yet in his glory days his name was seen as a mark of excellence.

Born in Oxfordshire in 1939, Symes had a difficult childhood. His mother was murdered in front of him, and he was put up for adoption. Dropping out of art school, he married at twenty-one and drifted into dealing, where he met and was greatly influenced by John Hewett. Symes worked on Mary Dark's stall at Portobello selling oak furniture, graduating to a first-floor shop on the corner of the King's Road and The Vale. Here he carried on dealing in oak furniture alongside medieval works of art and antiquities, including Egyptian objects, Roman glass and rings. He was already attracting customers like Simon Sainsbury.

Julian Barran remembers him staging an exhibition of David Hockney drawings in 1967. Everything changed for Symes that year, when Christo Michaelides walked into his shop. Described as 'A Greek God', Christo came from a wealthy family (but not shipping aristocracy) and brought a particular kind of glamour to the business. Shortly after meeting, Symes divorced and Michaelides separated from his girlfriend, and the two became close partners in every sense. With Christo's family money and cosmopolitan access, and Robin's eye, theirs was the most beneficial partnership in the London art trade, a case of two and two making ten. Known as 'the Symeses', they had charm and enjoyed parties, although Symes could also be aloof and even cutting. While Christo was an optimist, Symes had a melancholic, pessimistic temperament coupled to a supreme sense of self-belief, which may have been part of his undoing.

Together, Symes and Michaelides set up shop in Ormond Yard. Their gallery was described as a masterpiece, from the doorframes to the modernist furniture, and there were usually about a dozen works of art on display. A spiral staircase led upstairs to the library, where Symes would put out objects to admire on a long table. Danny Katz recalled going to the shop and seeing an inspiring mixture of Rennie Mackintosh furniture, Egon Schiele watercolours and Chinese cabinets, 'stylish beyond belief and it opened my eyes'. Symes had become an *arbiter elegantiarum.* The division of labour between the partners would be a matter of great contention later, with the Michaelides family wanting to prove that Christo was an equal partner in the business. It seems most likely that Symes was the better expert and salesman, while Christo ran the business.

Symes caught the zeitgeist, and his was one of the first galleries to brand itself with beautifully designed advertisements and minimalist typography. In 1971, he produced a catalogue entitled 'Ancient Art' to celebrate the opening of his new premises. This included thirty-three carefully chosen objects, mostly small-scale, from all over the ancient world, including a Cycladic harp player (2600–2000 BC) and a bronze head of Apollo (second to first century BC). Provenance was not given. In 1979, Symes came to the attention of the international press when the *New York Times* reported on his purchase of a Roman glass bowl at Sotheby's in London for $1.04 million.

In their heyday, no dealers lived more grandly. The couple knocked two houses together in Seymour Walk, Chelsea, and Katrin Bellinger recalled a swimming pool surrounded by statues and an art deco room filled with Eileen Gray furniture* It was a house of great elegance and exquisite food. Symes, who didn't

* This furniture became a great bone of contention later.

drive, was driven in a silver Rolls-Royce or a maroon Bentley. In New York, they lived in a former Rockefeller apartment converted by Philip Johnson. They also had an apartment in Athens and a summer house on the Greek island of Schinoussa in the Cyclades. Both showmen, the enduring image of them is in black tie.

Symes's clients were mostly in America, where matters of provenance would not be regarded as a serious problem until the 1990s. One of his best clients was Maurice Tempelsman, a Belgian-American diamond merchant (and walker to Jackie Onassis) whom he introduced to antiquities. According to Arthur Houghton, curator-in-charge of antiquities at the Getty Museum from 1982 to 1986, Symes suggested Tempelsman offer his collection to the museum to fund his divorce. Symes took Houghton to see the collection in New York and they agreed over a handshake on a price of a little over $13 million. Two thirds of that figure was for a fourth-century BC southern Italian marble carving of two griffins killing a fallen doe.[3] It later emerged that the sculpture had been found between 1976 and 1978 in a tomb near Ascoli Satriano, in Foggia, and passed from looters to Giacomo Medici, an Italian smuggler who controlled a team of '*tombaroli*', or tomb raiders. From Medici it went to Robert Hecht and then to Robin Symes, who sold it to Tempelsman.

From the mid-1980s, the Getty dominated Symes's business and would receive objects from the collectors he supplied. Typical were Lawrence and Barbara Fleischman, who filled their Upper East Side New York duplex with spectacular examples of classical art he sold them. Lawrence was a successful Detroit dealer in American art and, according to one account, he and his wife 'came under the spell of the dazzling presentations Symes staged for preferred customers in his private studio. There he often unveiled his best merchandise by sweeping aside a velvet curtain to reveal a dramatically lit object mounted in

18a. Mired in secrecy: the dismantling of the Pitt Rivers Museum in Farnham, Dorset.

18b. (*below*) Bruce Chatwin catalogues antiquities at Sotheby's.

19. Eminence grise: John Hewett.

20. (*above*) Portobello Market: the training ground for dealers.

21. (*left*) Queen of porcelain dealers: Pamela Klaber started at Portobello aged seventeen.

22. (*right*) Bennie Gray of antique market fame with the stolen head of Nelson.

23. Richard Green, the great survivor.

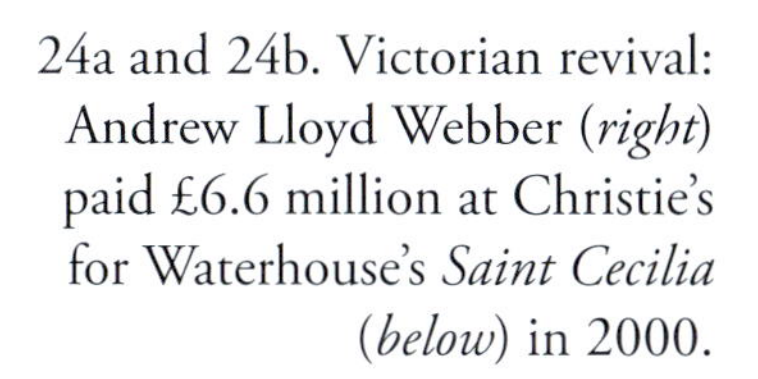

24a and 24b. Victorian revival: Andrew Lloyd Webber (*right*) paid £6.6 million at Christie's for Waterhouse's *Saint Cecilia* (*below*) in 2000.

25. (*right*) The first big sculpture price was £6.8 million paid for de Vries' *Dancing Faun* at Sotheby's in 1989.

26. (*left*) The 'Aphrodite' sold by Symes to the Getty for $18 million. The statue was later returned to Italy.

27. Double life: Robin Symes (*right*) and his partner Christo Michaelides sold looted antiquities to American collectors.

28. (*above*) The notorious Sevso silver hoard, now in the Hungarian National Museum.

29a and 29b. (*above*) Oliver Hoare and the Shahnameh manuscript he traded with the hard-line Iranian regime.

30a and 30b.
Mentmore Towers provided the most opulent house sale of the era in 1977.

31. The von Hirsch sale at Sotheby's, 1978 (*left*): the greatest sale brought into London in modern times.

32. Van Gogh's *Sunflowers* became the most expensive painting in the world when it sold for £24,750,000 at Christie's in 1987 (*above*).

33. Anthony d'Offay, whose gallery became an encyclopaedia of contemporary art.

34. The artists go it alone: *Freeze*, 1988.

35. YBA dealer in chief Jay Jopling with Damien Hirst.

36. Damien Hirst's shark defined London's new role as contemporary art city.

the centre of the showroom.'[4] In 1986, the Fleischmans gave and sold parts of their collection to the Getty, causing a stir in the archaeological world given the pieces' lack of provenance. In the same year, Marion True became curator of the Getty antiquities department. A former Boston curator specialising in Greek art, she would be at the centre of the storm over looted antiquities, already rumbling in the background.

True's predecessor, Arthur Houghton, had felt confident during his tenure that even if a source country claimed an unprovenanced object, they would have no proof of legality and, as he put it, 'In the mid-1980s curators of purchasing museums in the United States increasingly engaged in a game of don't-ask-don't-tell.'[5] Marion True later paid the price for this attitude, but at the time of her appointment she was no match for the salesmanship and charms of Symes and Michaelides, who lent her money alongside Lawrence Fleischman to buy a summer house on Paros.

Faced with the open chequebooks of such stellar clients in the USA, Symes's main problem was supply. The wars in the Middle East created opportunities for material to be spirited out of Lebanon, Iraq and Afghanistan. Typical was the Bull's Head excavated from the Temple of Eshmum in Sidon, Lebanon. The piece disappeared during the civil war and later reappeared on the black market. Symes sold it in 1996 for $1.2 million to the Colorado collectors Lynda and William Beierwaltes (to whom he sold over ninety pieces). It was identified by an archaeologist from a photograph illustrating an article in *House and Garden* two years later. The most sought-after material, however, was still to be found in southern Italy.

In 1988, Symes invited Marion True to his warehouse in Battersea to see a larger-than-life-size statue of a Greek goddess. Hugh Eakin described how True was bowled over by the work; she 'had never experienced a work of art in quite this way: "It was something that provoked you. You really felt this is what, when you stood in front of a cult statue, you were intended to feel." Fifth-century BC Greek sculpture was rare enough, but a work this fine, and finely preserved, was almost one of a kind. True consulted with several scholars; they shared her enthusiasm.'[6] The sculpture, erroneously believed to represent Aphrodite, was in three pieces (suspicious in itself, as this was a method used by smugglers to remove objects in portable sections). It had been found at Morgantina, Sicily, in the late 1970s, and illicitly shipped to Switzerland, where it was bought by Symes for $400,000. Dreaming up a 1930s Swiss provenance with forged documentation, he asked $24 million for the sculpture, making it the most expensive piece of ancient art on the market. With its obvious quality, questions were raised about why such an important item had never been published or exhibited before.

Marion True and the Getty trustees were only too aware of the problems of looting, but Symes still wore a badge of respectability. In her desire to acquire the sculpture, True allowed her critical faculties to be suspended. Overlooking the issue of provenance, she focused on her belief that it was by a Greek sculptor working in southern Italy and told the Getty trustees it was the 'single greatest piece of ancient art in our collection; it would be the greatest piece of classical sculpture in this country and any country outside of Greece and Great Britain'.[7] Hoping to circumnavigate the looting question, the museum resolved to use the statue as a test of their new policy of informing foreign governments to check their records before acquiring anything, and they sent photographs of the sculpture to the Italian culture ministry. Symes loaned the 'Aphrodite' to the Getty to be studied for a year, and it attracted a great deal of public interest. In 1988, True and Symes agreed a purchase price of $18 million. The trustees, however, were not entirely convinced by the provenance and demanded that Symes put up over $9 million in collateral during the period covered by the warranty, in case the statue turned out to be looted.[8] The sculpture would eventually be returned to Italy in 2007.

The fall of Robin Symes began in an unexpected way. In July 1999, he and Christo were guests at a dinner given by the collectors (and clients) Leon Levy and Shelby White, near Arezzo in Italy. Christo went to find his cigarettes and slipped on some steps. He died in Orvieto hospital the following day. Until this point, Symes had enjoyed an excellent relationship with Christo's family, but they were to be the main agents of his undoing. When they asked for their uncle's personal effects, Symes's response infuriated them, and they demanded half of the business assets in return. In Symes's view, 'Christo and I were partners, not in the business sense, but in the husband-and-wife sense. While we were both alive, we shared equally in the assets and profits and debts of the company, but after death they all passed to the survivor, to me.'[9]

Christo's family held a very different view, and his determined nephew, Dimitri Papadimitriou, hired Ludovic de Walden, an aggressive London lawyer, who rather enjoyed the art world's shenanigans. They also enlisted private detectives to follow Symes and demanded a share in the value of the Eileen Gray chairs he had sold. A master of concealment and deception, Symes would never agree that anything had ever been sold at the price that was suggested. De Walden managed to get an injunction to break into five of Symes's premises and seized all the documentation. Symes was followed in six countries and the legal costs are said to have amounted to $16 million.[10] Five locations turned out to be merely the beginning, as, according to de Walden, Symes stored his antiquities at thirty-three sites. It was reckoned that, across these spaces, there were 17,000 objects, with a stock value of £125 million.[11] Symes had underestimated his adversaries in Christo's family.

The furious relatives brought a civil case against him in London's High Court, at which Symes would pay dearly for his hubris. As one investigator put it, 'de Walden's relentless pursuit exposed Symes as a pathological liar, earning the dealer time in prison for perjury'.[12] What emerged was an astonishing story that focused on the nexus of looters in the Mediterranean, middlemen in Switzerland and Symes, who would 'wash' the items with fresh provenance like 'Left Egypt before the Suez war', before passing them on to American collectors. At the centre of the web was Giacomo Medici, who fed the looted items through various fences, including Symes, to auction rooms and galleries. Before his arrest in 1997, Medici shared a business address in Geneva with Symes, his best customer for more important items. His story is told in gory detail in Watson and Todeschini's book, *The Medici Conspiracy* (2007).

The case against Symes went backwards and forwards while he performed legal somersaults and continuous obfuscations. Nothing could be taken for granted. An Egyptian statue of Apollo seemingly sold to a Wyoming company for $1.6 million had in fact been sold to Sheikh Saud bin Mohammed Al-Thani of Qatar for $4.5 million. In March 2003, the Italian Ministry for Cultural Affairs announced that they had seized an extraordinary life-sized ivory head of Apollo, dated to the first century BC, from Symes. It was valued at $50 million and has the dubious distinction of being the world's most important looted antiquity.

Financially supported in the trial by Leon Levy, who died in 2003, Symes's own lawyers (he employed seven different firms) eventually caused his bankruptcy in March 2003, and he lost control of his business and his home. What began as a family civil litigation became a full-scale trial for fraud. The curious thing is that, in the end, Symes went to prison not for smuggling but for contempt of court. His friends said that he was deranged by Christo's death which 'affected his sanity', but his perjury and evasion suggested that he still had his wits, if not his judgement. Sentenced to two years imprisonment in January

2005 at Pentonville Prison, Symes was released after seven months. Even now his affairs have not been fully unravelled and objects from his storage continue to surface and be returned to Italy and Greece.

Chapter 24

The 'Sevso' Saga

*'May these, O Sevso, yours for many ages be, / Small vessels fit to serve your offspring worthily.'**

One morning in 1980, a few months after Peter Wilson had retired as chairman of Sotheby's and moved to France, he called his London assistant: 'You must keep this very secret. We have discovered this amazing Roman silver.'[1] The other half of 'we' was Rainer Zietz, the young German dealer based in London, whose speciality was medieval works of art. Zietz had been approached by Mansour Mokhtarzadeh, an art dealer in Davies Street, who had said, 'We have something very interesting to show you … We have it on consignment.' The object was a spectacular silver ewer. Mokhtarzadeh did not reveal the consignor, although it later turned out to be a Lebanese art dealer called Halim Korban.

Zietz contacted Wilson, who initially became the central figure in the episode. He also brought in Richard Falkiner, the former Christie's antiquities expert, to authenticate the ewer. Falkiner recalled: 'I was shown a silver ewer in excavated condition; it was in three pieces: the body, the handle and the base. It was immediately apparent to me that this fine object

* Inscribed in Latin on the largest and most elaborate 'Sevso' plate.

was not Sasanian but rather it was Roman, mid–late fourth century AD. It was one of the finest pieces of Roman plate and certainly the best that I had ever seen outside the amiable captivity of a museum.'[2] He percipiently suggested that 'It might be the beginning of more'. Wilson and Zietz bought the item through a firm called Art Consultancy Guernsey, an offshore company set up by Wilson's lawyer, Peter Mimpriss of Allen Overy, under the aegis of the Royal Bank of Canada.* Geoffrey Jenkinson, an antique arms and armour dealer, was appointed as the front man in Guernsey and the company paid £48,000 for the first piece.[3]

It was at this point that Mansour revealed the consignor to Zietz, who was able to go and meet Korban, whom he found to be 'a very pleasant, cultivated middle-class man'.[4] Korban told him 'It is not my treasure' and provided a story. According to his version, the hoard came from Lebanon, at the time in the middle of a civil war. Korban claimed that it had emerged during the digging for a petrol station and that there were more pieces, but he had not seen them.[5] The hoard was going to be sold one piece at a time by Korban's gallery in Geneva. Wilson and Zietz had bought the ewer on condition that they would have the right of first refusal for any further pieces that might turn up. As the items were dribbled out over the next few years, they were more magnificent and more expensive each time. The most spectacular piece was the enormous 'Sevso' plate, embellished with elaborate scenes of hunting and country pursuits, which bore the Latin inscription addressed to the owner: 'May these, O Sevso, yours for many ages be, / Small vessels fit to serve your offspring worthily'. The inscription led to the conjecture that the hoard was made for a Roman general who buried it on being forced to retreat. The plate cost £225,000.

* Zietz negotiated a one-third share of the profit on all the pieces.

Wilson and Zietz were dazzled by the unexpected magnificence of the 'Sevso' plate, and fully realised the importance of buying any further items that emerged to keep the group intact and save it from being sold piecemeal. At the same time, Mimpriss advised that, on account of the high value of the pieces, Korban should provide an export licence for every item. As these were in Arabic, it was also agreed that each licence should be countersigned by the Lebanese Embassy in London, with an appropriate authentication to be handed over directly to Allen & Overy. On the strength of these documents, Wilson and Zietz bought the 'Sevso' plate in December 1980, and in the course of 1981 and 1982 they bought six more equally impressive items. They were confident that, with just these eight items, the obvious potential owner of the group was the Getty. Wilson realised that selling the pieces was always going to be a private sale rather than an auction, because even in the undeveloped regulatory state of the antiquities market at the time, the silver was without clear history and provenance. He saw no reason at this time, however, to suppose that the hoard was unsaleable.

By 1982, Peter Wilson, who had provided the investment, was running out of money. Mimpriss therefore proposed bringing in another of his clients, the Marquess of Northampton, who, after going to see the items in Zurich, was keen to participate in the venture. He financed four of the eight items so far acquired. Despite the separate ownership, it was agreed that the hoard should only be sold in its entirety.

In 1984, it was decided to offer the eight pieces of treasure to the Getty Museum, who were interested, subject to the objects being sent to them for further inspection before making a final decision. The Getty believed the owner to be Geoffrey Jenkinson, who asked 26 million Swiss francs, or just over $12.6 million, for the eight pieces so far revealed. Arthur

Houghton, the curator of the department of Antiquities, who was fluent in Arabic, had misgivings about the objects' export licences. The museum asked the US Embassy in Beirut to look into the matter. When they got in touch with the relevant official, it was clear that he had nothing to do with any of the documents and knew nothing about the treasure or its export. When Jenkinson was confronted with this fact at a meeting with Getty officials at Allen & Overy's offices in London, he became 'incensed and outraged'. He announced that he had no choice but to rescind his original offer and, after a pause, reoffered the objects at $25 million, double the original price.[6] To Houghton, the whole deal smelt fishy, and the new price was, in his opinion, far too high.

Peter Wilson died on 3 June 1984, and without his astute attention, the 'Sevso' deal ran into difficulties. As Geraldine Norman put it, 'Wilson could step sure-footed as a goat through the rocks and precipices of international law, international smuggling, numbered accounts and offshore trusts, but his heirs, his friends, and his legal adviser did not have the agility to follow in his footsteps. Very quickly, they found that the central problem in marketing the treasure lay in its origin.'[7] The treasure was sent back to England, and Houghton discovered that the main shareholder was the Marquess of Northampton. Faced with the false export licences, Mimpriss, without Northampton's knowledge, paid Lebanon's director of antiquities £628,000 to 'rectify' the export documents or provide new ones. Northampton, now the prime mover, bought four more pieces in 1987 and, around the same time, he gained ownership of the entire hoard by buying out Wilson's heirs. When all the pieces were finally revealed, the 'Sevso' treasure contained fourteen silver vessels and plates and the copper cauldron that had allegedly contained them since antiquity. Geraldine Norman thought that, 'It may come to be seen as the greatest ancient treasure ever discovered.'[8] In fact, it became the most notorious.

Having failed with the Getty, Northampton decided to sell the treasure via Sotheby's in New York, where it went on exhibition in February 1990, its first emergence on public view. The catalogue stated that it had been found in what was once the province of Phoenicia in the Eastern Roman Empire. There was at first great excitement, until the story of the false export licences emerged and the new export documentation was also declared to be false. The sale was cancelled. Northampton sued Peter Mimpriss and Allen & Overy for damages caused by fraud in relation to the acquisition of the silver. Allen & Overy settled out of court at the cost of Northampton's entire investment of £15 million.

The hoard was now claimed by three countries: Hungary, Yugoslavia and Lebanon. Lord Northampton found himself being sued for possession by all three in a New York court. By 2011, the only country which was still actively claiming ownership through the media was Hungary, and the director of the Hungarian National Museum, Dr László Baán, appeared on television making his claims and lamenting the fact that the silver was not available to scholars. He contacted Northampton and explained that Hungary could not buy something that 'already belonged to them'. However, they could reimburse the costs incurred, which amounted to what he had paid for the items.

The Hungarians believed that the silver had been discovered in around 1975–6, by a young man, József Sümegh, in Polgárdi, near Lake Balaton. Sümegh had, according to this account, taken it piece by piece to Budapest, where he launched it onto the market. He was, however, soon discovered in the outbuilding where he had stored the silver, hanging by his neck. He had evidently been murdered, and the silver had gone. Hungary bought the first seven pieces of the silver from Lord Northampton in

2014, and the rest in 2017. Today, the 'Sevso' treasure greets visitors on arrival at the National Museum of Hungary. Due to the monumental size of the individual pieces, the collection can be considered the most important find of Roman silver ever.

Chapter 25

The Most Improbable Deal: 'Go see Oliver Hoare'[1]

Stuart Cary Welch to Arthur Houghton

Of all the deals described in this book, one stands out for sheer oddity. It was the deal struck by Oliver Hoare (1945–2018), one of the most colourful figures in the London art market, with the hard-line Iranian regime. The deal seemed inherently impossible, and only somebody with Hoare's chutzpah could have dreamt it up and followed it through. The dealer was lucky in every way, because, within a month of completing this 'deal of the century', he achieved fame for all the wrong reasons, being engulfed in a scandal which would certainly have doomed the project. Before discussing the details of the story, it is worth outlining the background of the Islamic art market in London.

In London, Islamic art attracted little attention until the early 1970s, with the appearance of the Iranians, followed by the Gulf Arabs a decade later. The main Iranian dealers were Mehdi Mahboubian in Brook Street (and New York), selling Near Eastern antiquities and glass, and Mansour Mokhtarzadeh in Davies Street, who had approached Rainer Zietz to sell the first 'Sevso' ewer. American museums, particularly the Metropolitan

Museum in New York, were leading the Islamic market, while the Getty focused more on Greco-Roman objects. The 1980s saw the development of a niche community of Islamic collectors including Edmund de Unger, David Khalili, the Sultan of Brunei Sheikh, Nasser Al-Sabah Sheikh of Kuwait and Arthur Sackler. There were no remarkable sales at Sotheby's or Christie's. Instead, the watershed was *The Arts of Islam*, a 1976 exhibition at the Hayward Gallery and it is a story more dominated by collectors than dealers.

The Gulf buyers came to London with Sheikh Nasser Al-Sabah of Kuwait. He and his wife, Sheikha Hussa Al-Sabah, were collectors of great discrimination, who set about buying objects for their museum in Kuwait. The collection was formed over time mostly from items coming up at auction. By the new millennium, there were various national projects to build Islamic museums across the Gulf, the most conspicuous being in Doha. The biggest buyers in the late 1990s were two compulsive collectors, Sheikh Saud bin Mohammed Al-Thani of Qatar and his cousin, Sheikh Hamad bin Abdullah Al-Thani, who formed major collections which included Mughal works of art.

The most important American collector of Indian and Islamic art was Stuart Cary Welch, based at the Fogg Art Museum, Boston.* Cary Welch had an impeccable record as a collector, academic and consultant to – amongst others – the Metropolitan Museum of Art. His main competition was Prince Sadruddin Aga Khan, uncle of the present Aga Khan, whose collection would eventually form the foundation of the Aga Khan Museum in Toronto. The museum was initially planned to be in London, but encountered such difficulties with the planners that he was forced to abandon the idea and go to

* When his collection came to Sotheby's London in 2011, it made two sales: 'The Arts of Islamic World' and 'The Arts of India'. The estimate was for £5–7 million, but the sales made £29.3 million.

Canada instead. Together, these collectors created a demand for masterpieces which pushed up prices.

When Arthur Houghton senior died in 1990, his son, Arthur Jr, decided to sell his extraordinary sixteenth-century copy of the *Shahnameh*, the Persian national epic. The Houghton manuscript is the richest version of this *Epic of the Kings*, containing 759 folios, of which 258 are miniature paintings. In 1976, seven pages from the manuscript had been sold by Houghton's father at Christie's for £785,000. Houghton Jr wondered to whom he should speak, and whether he should offer it to Sotheby's or Christie's, although he was sure that he did not want a further piecemeal dispersal. He rang up Cary Welch for advice. 'Oliver Hoare', came the clear answer, 'Go see Oliver Hoare.'[2] Hoare belonged to an old English banking family, educated at Eton and then the Sorbonne, but he remained rather mysterious. It was believed that both his parents worked for MI6. Fascinated by all things Persian, he had gained introductions to the collectors in Iran and opened a gallery with what one collector described as 'breathtaking mark-ups'. Beloved by women who could not resist his refinement, charm and good looks, Hoare married Diane de Waldner, a French oil heiress, in 1986, and through her he gained access to the European aristocracy.

In April 1991, Houghton visited Hoare's gallery in Victoria: 'The gallery itself was darkened as I entered. Heavy velvet curtains covered the windows. There was a faint scent of incense in the air. It was all theatre, of course, but it was difficult for me not to succumb, feeling I had entered one of the Oriental salons that I had known in Beirut.'[3] Six months later, they met again at Hoare's house in Chelsea, and the dealer told him that he had been asked by the Iranians if he had any important work of art that could be repatriated to their country. One of the problems was how any such object would be paid for since Iran was emerging from a disastrous war, had no money and parliament would never allow money to be spent on buying art.

If the *Shahnameh* was to be acquired, they would have to find a means of paying for it off-budget. Above all, the sale had to be secret and scandal-free. After a lot of toing and froing, Hoare hit upon an ingenious idea: 'Since I'd heard the Museum of Contemporary Art was closed, I suggested that there might be things there that didn't fulfill a role in their cultural plans.'[4] The Iranians agreed to a swap of the *Shahnameh* for Willem de Kooning's *Woman III*, which had been acquired by the shah's wife, Farah Pahlavi, when she founded the Tehran Museum of Contemporary Art. But were these two very different items of similar value? Hoare and his Iranian counterparts came up with an inventive solution. They established an artificially high equal value for each work. That allowed both sides the fiction that they had not given away a thing, while making a bargain and avoiding criticism when the deal became publicly known.[5]

Arthur Houghton describes what happened next: 'On July 27 [1994], *The Shahnameh*, in three heavy cardboard boxes, left London for Vienna, where it was stored overnight in a van in the international zone of Vienna airport. The following day, an unmarked Iranian 727 arrived in Vienna, the van with the boxes enclosing *The Shahnameh* folios was attached with chains to the aircraft.'[6] The de Kooning painting was flown from Tehran on the same plane and taken to the Amman Gallery in Zurich, where it was subsequently sold to the American collector David Geffen, for $20 million.

The sale was timely, for a month later, Hoare would be at the centre of a press storm when the London tabloids exploded with the story of his affair with Diana, Princess of Wales. He had detached himself from the relationship a year before, but it appears that she had become fixated on him, making hundreds of phone calls to his house from Kensington Palace, all supposedly recorded by the Metropolitan Police. Somebody within the palace had leaked the story, which would have derailed the deal. Had it all been known before, there is no question that the Iranians would have aborted. As it was, following the revelations Oliver Hoare tried to keep a low profile, with little hope of success.

The resultant publicity did not affect Hoare's business, and he had gained the trust of Sheikh Saud Al-Thani, the biggest buyer in the market. As chairman of the National Council for Arts, Culture and Heritage for Qatar, the sheikh was impervious to high prices and had a sure instinct for quality. Hoare once took him to Eskenazi's shop and, although he knew nothing about Chinese art, he instinctively picked the best pieces. In 2005, however, Hoare became embroiled in an inquiry into allegations that Sheikh Saud had embezzled funds when it emerged that he provided two invoices, the real one and an inflated one for the museum. Hoare maintained his innocence, as did the sheikh, and charges were eventually dropped. Hoare's legacy will always be the astonishing deal with the Iranians, which others have tried to reproduce without success.

Chapter 26

Game Over Bond Street

'The era of great galleries, of clients visiting, sweeping into the lush premises was beginning to wane.'[1]

Alex Wengraf

Sometime in 1992, the chairmen of two Bond Street giants, Asprey's and Sotheby's, met. John Asprey asked to see Lord Gowrie at Sotheby's on what he called a neighbourly mission. Such were the closures caused by the recession, and so great were the number of boarded-up shops in London's faubourg, that Asprey proposed establishing a fund to put flowers in the empty windows to disguise the extent of the exodus. For some time, art dealers had been deserting the street with its stratospheric rents. Those who were not already occupying higher floors were moving to smaller premises south of Piccadilly, around Duke Street. It even looked as if Bond Street might be eclipsed by Sloane Street as the shopper's destination of choice, but what nobody could then envisage was the arrival of the fashion palazzo to replace the art palazzo. The changing face of Bond Street became a metaphor for the changing face of traditional art dealing. A generation had retired, not to be replaced.

The disappearance of so many Bond Street galleries was a gradual affair, hastened by the 1990s recession. It affected the whole of the West End, and many Bond Street fixtures closed over the next two decades: Frost and Reed for sporting art, Leger for British eighteenth century, the Fine Art Society for Victorian paintings, and the furniture dealers, Mallett's and Partridge. Next door in Bruton Street, the Lefevre Gallery prospered until they too were closed by the impact of 9/11, their place in London's impressionist scene overtaken by the Nahmad family, who operated from warehouses in Switzerland. The closures also affected St James's, south of Piccadilly: the Heim Gallery, Leggatt's and Spink. The most dramatic closure was that of Agnew's in the new millennium. The exception to this trend was Richard Green, who successfully sailed on, opening more galleries on Bond Street and picking up clients and artists as the other firms closed.

Compounded by the recession, there were many forces at work, starting with the steep rise in rents for street-fronted spaces on Bond Street. This encouraged the appearance of the art agent with lower overheads, little or no stock and greater flexibility. Alongside this, the internet radically changed the nature of dealing. It was no longer necessary to have a large shop if you could send likely buyers a high-resolution image. The internet removed much of the secrecy of the business, making it possible for anybody to look up what had been paid for an item at auction. It also vastly increased the potential sources of purchasing, chipping away at the monopoly of London's West End. Of course, a client who might be intrigued by an image would still want to see the painting, but this could be done from a second-floor space in Jermyn Street as easily as from a shop front in Bond Street. The rise of international art fairs was another important factor in the decline of gallery visitors, especially after 9/11. Above all, tastes were changing. The generation that got rich as a result of 'Big Bang' deregulation in 1986 was more attracted to Damien Hirst than Thomas Gainsborough.

If you had asked anyone in the 1980s which was the grandest and most stable gallery in London, the chances are that they would have mentioned Agnew's. The seemingly indestructible battleship was symbolically moored at the lower end of Old Bond Street, halfway between Sotheby's and Christie's. As the firm entered the 1990s, the façade still looked impregnable, but the seeds of its destruction were already sown. During the 1960s, Agnew's had had the best reservoir of clients in the business, for whom they bought and sold great paintings from their grand premises. With the rise of auction houses and the willingness to consign major old masters to auction, the firm had been forced to adapt. This meant becoming the chosen bidder at auction on behalf of American museums. Although prestigious, as they could still claim to have handled the works, this activity brought little profit and did not help cover the gallery's vast overheads, which included five porters and a framer. Such fixed costs would be one of the many factors that made Agnew's vulnerable to art agents and more nimble operators without expensive stock.

During the 1980s, an ambitious young agent, Robert Holden (1956–2014), invented a new genre of dealing, the sole purpose of which was to negotiate sellers' commissions by putting Sotheby's and Christie's in competition. His earliest coup was while still a pupil at Eton, when he bought a piece of silver for £10 in a house charity sale and resold it in the High Street later the same day for £200. When asked if he would contribute some of his profit to the charity, he politely refused.[2] Agnew's had felt confident in their position as the London art trade's arbiter and ringmaster until one of its traditional clients asked Holden to negotiate the sale of Wright of Derby's *Mr and Mrs Thomas Coltman*. Holden, who had nerves of steel and a touch of James Bond about him, had become involved as a friend of the owner's daughter. His strategy was simple: negotiate the best deal from Sotheby's and Christie's and take a cut from the seller. With no special expertise, he nevertheless revealed how flexible auction house commissions could be.

Sold in 1984 for £1,404,000, *Mr and Mrs Thomas Coltman*, now in the London National Gallery, upset the status quo, and Holden became the go-to man. Clients who had previously defined themselves as Christie's or Agnew's now felt that, by consulting Holden, they might get a better deal. If Agnew's was probably the more damaged by this development in the long term, in the short term, Christie's were more vexed by Holden's operations. Hugh Roberts, head of Christie's Furniture Department, recalled that 'my heart sank when I heard he was getting involved'.[3] Even Richard Green, who saw every change as an opportunity, felt that the arrival of agents like Holden was 'not that helpful to dealers'.[4]

Holden's clients tended to be the typical country house owner who was too gentlemanly or too inexperienced to negotiate terms themselves. His mediation often gave Sotheby's a chance at business they might not otherwise have been offered. Typical was a Midlands house sale, which presented Holden with a conundrum. The family had always been Christie's clients and saw Holden's role as a way of coaxing the best terms out of them. If that meant consulting Sotheby's, so be it. When the two proposals arrived, Holden had to admit that the Sotheby's proposal was better, but the family insisted on staying with Christie's. George Bailey, the Sotheby's managing director, pointed out to Holden, 'If you go along with this you are not independent, and we will no longer play this game.' Realising that his business model was in jeopardy, Holden was forced to consider whether to persuade the family to accept Sotheby's proposal or to resign. The next day, he made the most difficult call of his career to Christie's, having already told them that they had the sale, to say that they did not have it after all. They were understandably furious, but Holden lived to fight another day. He served his clients well and the property he handled sold. This would not always be the case with the agents who followed him.

When Sir Geoffrey Agnew died in 1986, something went out

of the firm. His successors did not have his chutzpah, although they still had some notable successes. The following year, Sotheby's Chester offered a painting on slate as a nineteenth-century copy, *Portrait of a Pope*. It was bought by Buffy 'the hoover' Parker for £180. He was a genial 'numbers' dealer who would buy twenty pictures and put them into Christie's South Kensington on the grounds that five might make a profit. He took the picture to Christie's, where it was identified as Pope Clement VII by Sebastiano del Piombo. The painting sold for £320,000 to Agnew's, who resold it to the Getty for $10 million in 1991. The internet would soon render such bargains less likely. Three years earlier, the Getty had used Agnew's to bid for them to acquire Pontormo's *Halberdier* at Christie's New York for a record $35.2 million, then the highest price for an old master painting sold at auction.

Agnew's traditional model was about holding great stock, something increasingly expensive in the new climate. 'The trouble,' as Christopher Kingzett saw it:

> was that Agnew's went on believing they were right, and others were wrong. It was a model that could survive until you bought a very expensive picture that you couldn't sell at the price you hoped to achieve. The picture that paralysed Agnew's in the 1990s was a Rubens *Holy Family*, bought privately in South America no doubt for too much money. Agnew's asked $20 million for it. When the painting failed to sell, instead of taking a quick loss they held onto it, increasingly desperate.[5]

At this critical point another disrupter appeared, developing the Holden agent model for the internet age. Simon Dickinson, for whom, as far as Agnew's were concerned, 'the rules did not apply',[6] is one of those trade experts with a brilliant eye and visual memory, though untutored in academic art history. He left Christie's in 1993, knowing where every old master in Britain was to be found, having valued most of them. Dickinson had observed that 92 per cent of old masters sold at Christie's went to the trade and saw an opportunity to match the seller with the end buyers.[7] He joined up with a paintings dealer, David Ker, and wrote to hundreds of owners, offering to sell their paintings. Lord Gage, the owner of Firle, offered him his first chance, a Correggio *Head of Christ*, which Dickinson sold to the Getty.

Dickinson used to advertise in art magazines as 'Sotheby's, Christie's or Dickinson', and one former colleague described him as an incurable optimist with extraordinary self-belief. He is a buccaneer who gained a reputation for offering to get very high prices for pictures on consignment, much more than Agnew's paid buying them for stock, or the auction houses giving a realistic estimate. The business model that evolved was fundamentally internet-driven and was made possible by the advent of high-resolution images. Dickinson's *modus operandi* was to contact forty of the great collectors of the world by email, attaching an image and hoping that somebody might then come and see the painting. There were casualties in this approach, which left paintings unsaleable after they had been turned down by the top collectors (in trade parlance, 'burnt'), but Dickinson had enough successes to create a striking and fashionable business that made it harder for traditional dealerships like Agnew's to compete.

Dickinson is one of the most talked-about figures in the London art trade. This is partly because of the spectacular discoveries he has made, most famously, Botticelli's *The Virgin Adoring the Christ Child*, then thought to be a studio work, but now fully accepted and in the Scottish National Galleries. He also has a reputation for embroiling himself in legal disputes to the extent that some have wondered whether he was simply blithely unaware of the rules. In 2007, he made his biggest deal to date, negotiating the sale of the Northbrook collection of

Dutch paintings to Prince Hans Adam of Liechtenstein for £17 million. After the sale, he was arrested for supplying incomplete or false information to the Export Reviewing Committee. The essence of the case was the alleged manipulation of figures to clear the export of Michiel van Musscher's *Portrait of an Artist*. The case was allegedly dropped because the British Treasury and Foreign Office were negotiating a new tax treaty with Prince Hans Adam, which would have been jeopardised had the prosecution proceeded. Sir Ronald Grierson, a trustee for the prince, negotiated the settlement.

Against this backdrop, Richard Green defied gravity by surviving with the old-fashioned model of a window dealer and proving that every snare was an opportunity. Until the mid-1980s, Modern British, known colloquially as Mod Brit, had been a primary market in Cork Street, until Green began to develop it as a secondary market. It had previously appealed to a small band of artistically minded collectors and connoisseurs, but Green's interest in Moore, Hepworth and Nicholson put a new category of clients in front of their art and made it mainstream. This was no mean feat, given that selling abstract art to Yorkshire farmers was quite an achievement, and Green came to dominate the market for artists like Ivon Hitchens, and produced catalogues for shows of Lowry and Heron amongst others. As a saleroom specialist commented, 'Green was the only person at the high end of the Mod Brit market buying largely for stock where most of the others were buying on commission.'[8]

Once the internet made prices a matter of common knowledge, many dealers were put off buying at auction, as their mark-up became widely known. This had no impact on Green, who thought it was not the price paid that was important, but the value of the painting. If a client pointed out what a painting had made at auction, he would simply respond that he had bought it undervalued and was now offering it at its true value. Green worked extraordinarily hard, eschewed smart art-world parties, lunches and private views, and would quietly go around the salerooms at the weekend, when he could be alone and unobserved. As one employee commented, he is 100 per cent involved twenty-four hours a day, and his only relaxation is horse racing. There is a story that, one day, a member of staff asked Richard what his favourite picture was. Not given to bonhomie, he answered 'a sold one!', and gave a thin smile.

In 1995, Godfrey Barker, the *Daily Telegraph* saleroom correspondent, published the accounts of the top ten dealerships in and around Bond Street during the recession of 1993–4.[9] Richard Green was riding the storm with the highest turnover at over £38 million, against Agnew's £16 million; Green made a £1.15 million profit, against an Agnew's £2.85 million loss; Green held fresh stock worth £31.5 million while Agnew's held £20 million of perhaps older stock. Green paid himself £325,000 a year, but the remuneration of Agnew's directors did not figure in the top ten dealerships.

Agnew's would continue until 2008 when the firm closed after 195 years. The final nail in the coffin was their splendid Bond Street gallery. Its value had risen out of all proportion to the business. The crux came when it required millions of pounds to renovate, and the family shareholders decided to realise the value of the building instead. Julian Agnew, the sixth-generation chairman of the firm, told the *Antiques Trade Gazette*, 'We are neither the Tesco's of the market like the big auction house nor a small dealer with only an assistant and a dog who can run on very low overheads.'[10] He was spot on with his next remark: 'Changes in the market and technology make a gallery no longer necessary unless perhaps you are a big contemporary art business.'[11]

*

Before contemporary art swept all before it, the favoured collecting habit of the very rich had been impressionism and post-impressionism. Throughout the 1970s and 1980s, the market-makers in this area were the auction houses. There were two major London dealerships at the highest level, Lefevre and Thomas Gibson, both just off Bond Street in Bruton Street, who offered the only real competition to Sotheby's and Christie's. Lefevre had a distinguished history, and their continued success was the result of Desmond Corcoran's purchasing and Martin Summers's selling skills.[12] They were the most exclusive dealership in the capital, and their first-floor showroom was described as 'the art collector's equivalent to the Connaught Hotel'.[13] It was said of Summers's charmed existence, 'In life you have to take the smooth with the smooth.'

Lefevre depended on rich foreign customers from Greece and North and South America. By 1971, they had their first million-dollar picture, a Van Gogh *Still Life*. Very soon afterwards, they sold the Chicago collector Leigh Block's Van Gogh *Self-Portrait* to Niarchos for $2.5 million. They even sold a collection of Degas bronzes to Norton Simon.* During the 1970s, their biggest client was Carlos Pedro Blaquier, the Argentinian sugar baron who made a fortune selling top-grade sugar to Coca-Cola. Lefevre sold him nineteen impressionist paintings, including Lord Rothschild's Cézanne *Harlequin* for $2 million. This happy relationship ended with the Falklands War in 1982, but 9/11 was the real game-changer as Americans stopped coming to London. This created a loss of confidence across the trade, and the partners decided to close the gallery.

Thomas Gibson was more in the half-agent half-owner model of operating, and he brought the scene into the twentieth

* In 1972, they bought an entire collection of Degas bronzes from the widow's estate, which included the Parisian casting firm Hebrard, for $1 million. They are now in the Norton Simon Museum in Pasadena, California.

century with artists like Giacometti, when Lefevre were still selling classic impressionism. The firms did business together and frequently worked with international colleagues, notably the New York dealers Bill Acquavella (the greatest dealmaker of the age) and Eugene Thaw. They also worked with two European dealers, Ernst Beyeler from Basel and Heinz Berggruen. These two brilliant dealer-collectors were often in the background of big London transactions.

Gibson had trained at Marlborough Gallery in New York, returning to England in 1969 to set up his own dealership. He was encouraged by Ernst Beyeler after he put himself on a plane to Basel next to Beyeler. Gibson had £15,000 in cash and a £15,000 overdraft facility. With this he bought a £30,000 Leger charcoal from Beyeler, adding, 'Please consign me these six other pictures' chosen from the dealer's stock. 'Can you sell them?' Beyeler asked. Gibson succeeded in selling the works and they split the proceeds 50/50. The Leger is now in the Royal Ontario Museum, Toronto. Gibson had learnt from Marlborough how to ask high prices and justify them. With the Livanos and Embiricos families as clients and the Earl of Gowrie cutting his teeth as his gallery salesman, Gibson was known for having a small but extremely discriminating and expensive stock.

Gibson would ring up one of the auction house impressionist departments and say: 'Whenever I am bored I either like to buy something or sell something.' When Christopher Davidge wanted to ginger up Christie's Impressionist Department, he realised they needed Gibson's firepower and asked him to be a consultant. The arrangement made the latter a lot of money and brought the department up. The practice of auction houses giving guarantees made it harder for dealers like Gibson to compete. Today he has grand old man status, and the gallery continues under his son, Hugh Gibson.

If Lefevre and Gibson were old school, the dealers who kept the London impressionist market moving were the Nahmad family. Simon de Pury described them as 'the only bidders during the worst times who thrived like vampires on market bloodlettings like the nineties depression'.[14] According to Michel Strauss, 'The Nahmad family, Jewish merchants emanating from the Middle East, have had the greatest influence on impressionist and modern art auctions in London, New York and Paris.'[15] Heirs to a Syrian banking fortune, their first venture into the art world was in Milan, before they opened galleries in New York and London. The Nahmads bought most of their stock at auction and stored it for long periods at the Geneva Freeport. As Strauss recalled, the family would turn up at auction *en masse*, occupying as many as eight front-row seats, from where they discreetly signalled their intentions to the auctioneer. They bought in the middle and top range, and usually sold only when the market was right. An example of a masterpiece they handled was Monet's *Le Pont de Chemin de Fer à Argenteuil*, which they bought for £6.2 million at Sotheby's London in 1988 and sold twenty years later in New York for $37 million. In Strauss's opinion, 'There isn't any dealer in the post-war years who has transformed the auction business more than the Nahmad family.'[16]

An Impressionist Department director estimated that the Nahmads bought between 15 per cent and 20 per cent of every major sale in London, building up an enormous stock. Strauss recalled that one of the family, Joe Nahmad, 'specialized in coming up to the auctioneer at the end of a sale to try to mop up cheaply any unsold lots'.[17] They tended to have a very specific shopping list: Utrillo, Renoir, Vlaminck and, later, Picasso, Matisse and Chagall. A member of the trade recalls inspecting their space at the Freeport in Geneva in the early 1990s, estimating that he saw over £400 million worth of art. The Nahmads caused headaches for the financial officers at Sotheby's and Christie's with their payment arrangements – debts were usually paid with pictures for sale instead. This suited the departments, and if major sales were looking a bit thin before the catalogue deadline, there was a tendency by both Sotheby's and Christie's to go to the Nahmad warehouse to select eight or nine items, even if they were recently purchased and, in some cases, had a very low chance of sale. The family did not open a gallery in London until after the turn of the millennium.

One casualty of the 1990s was Spink, the art world's department store. When the owners tried to sell it for £42 million in what *Private Eye* called 'the riddle of the Spinks', the deal was kicked into touch by the 1990 invasion of Kuwait. Christie's, their next-door neighbours, snapped up the company for the real estate. Christopher Wood thought that 'Christie's had to buy Spink's business, which they really did not want, to get the building'.[18] They set about dismantling it and attempting to absorb what they could, but most Spink experts left to start up their own businesses.

Under David Posnett's energetic leadership, Leger had recovered from the Keating scandal (see Chapter 16) and become a major player in the British paintings market. Posnett could be abrasive but was well respected and made some spectacular discoveries, notably Edward Haytley's *Elizabeth Montagu and her Family*, William Hodges's *Portrait of Captain James Cook* and John Singleton Copley's *The Fountaine Family*. He made something of a speciality of George Romney. With the death of Paul Mellon in 1999, the taste for British art, especially sporting art, withered and with it many of the dealers that brought character to Bond Street. Posnett had already sold the firm to Christie's, who created Spink-Leger as a vehicle to do private sales at a time when auction houses were tentatively stepping into this area but did not yet have the experience. They wrongly believed they needed to buy dealerships to assuage the art trade's hostility to this development.

Sotheby's had similar ambitions to acquire a retail gallery outlet and bought Robert Noortman's gallery of Dutch paintings. He had opened in London in 1974, while still based in his hometown of Maastricht. If his first great disruption to the London trade was setting up the Maastricht Fair, then his second was selling his gallery to Sotheby's in 2006, thereby assisting the shift by which auction houses became art dealers. Noortman operated at all levels from Rembrandt downwards. One dealer said that 'he had an ear rather than an eye', and there was some truth in this, because he seemed to be everywhere and all knowing. A superb salesman, Noortman appealed to Sotheby's, who hoped that they were buying his charisma, although they were sorely disappointed as he died the following year. His reputation suffered in light of revelations concerning missing pictures that emerged after his death.

The most conspicuous dealer in Dutch paintings in London was Johnny van Haeften, for whom art dealing and clients were treated like a most enjoyable hobby. 'Boyish, blonde and jovial with the comfortably corpulent physique of a *bon viveur* who uses his charm as a bait',[19] he was the archetypical Duke Street operator, buying paintings in shares, dispensing hospitality and very popular with his colleagues. Van Haeften trained at Christie's, where he set up the stamp department, a good training for Dutch paintings, which, with their characteristic minutiae, he characterised as 'overgrown stamps'. He set up a gallery with his wife Sarah, starting from home in Fulham: 'I had no clients, no knowledge, and it was a leap of faith, but I started making a card index of everybody we hadver met.'[20]

Of all the dealers in London, van Haeften is one of the most sociable, remarking that 'All my clients became friends'.

One evening, the ballet dancer Rudolf Nureyev came to van Haeften's gallery just as he was closing. The dealer welcomed him in Russian, to which Nureyev responded, 'Don't ever speak to me in Russian, only French or English.' Van Haeften then asked, 'Would you like some vodka?', to which Nureyev replied, 'Don't ever offer me vodka, I'm teetotal!' Undeterred, van Haeften said, 'I think we had better start again. Good evening Mr Nureyev, would you like a cup of tea?' and they became friends.[21] One of van Haeften's more extraordinary stories was when somebody brought in a photograph of a Pieter Brueghel the Younger which had been in the family, it was said, since it was painted. The most unexpected part of the story was that it was located out in the bush, three hours from Nairobi, Kenya. Van Haeften flew to Kenya, drove to a 1930s villa and there was Breughel's *Census at Bethlehem*, acquired on a diplomatic mission in Antwerp in 1611, with unbroken provenance down an English family. He bought a half share in the picture to give the family the money they needed immediately and took it to London, where it was cleaned beautifully and sold on the second day of the Frieze Fair. Today, van Haeften reckons he has sold over 5,000 paintings, 100 of them by the Brueghel family. After 9/11, the passing trade in Duke Street began to fall apart, and when the entire building was sold, he closed the gallery to operate from home in Richmond.

Many traditional old master dealers survived the 1990s, but very few maintain a ground-floor shop window. Although Colnaghi passed from one owner to another, in Christopher Wood's view it began its long decline with the retirement of Jim Byam Shaw. As Wood put it, 'the name survives, and now belongs to a group of dealers occupying a small space behind the wonderful old gallery, of which only a small part survives'.[22] The Byam Shaw/Thesiger tradition is kept alive less at Colnaghi today than at Patrick Matthiesen's gallery.* Matthiesen has made something of a speciality of scholarly catalogues, publishing over sixty to date, each a monument to its subject. One particularly memorable example accompanied his 1984 exhibition *From Borso to Cesare d'Este: The School of Ferrara 1450–1628*.

The production of scholarly dealers' catalogues was relatively new, and pioneered by the Heim Gallery. As Matthiesen recalled, 'In the 1970s, most old master catalogues consisted of one or two laboriously produced colour plates, a handful of black-and-white illustrations and the artist's name, dates, the sizes of the canvas and possibly a line or two suggesting a dating or an opinion.'[23] In 1981, Matthiesen moved into his new galleries at 78 Mason's Yard, the first art gallery to be constructed in London for almost seventy years. He marked it with an exhibition in aid of the Naples earthquake appeal, *Italian Baroque Painting 1600 to 1700*, which included Orazio Gentileschi's *Madonna and Child*. Occasionally, he would leave his baroque niche and exhibit early Italian or eighteenth-century artists like Louis Gauffier.

Hazlitt continued after Jack Baer's retirement under the leadership of John Morton Morris. Morton Morris has an invaluable network and has made important discoveries, including the Pieter Saenredam he recognised at Bonhams, and a Rembrandt *Self-Portrait* he found in an auction in Gloucestershire, which is now at the Getty. His partner, Michael Simpson has one of the best eyes in the business. A slight air of mystery surrounds John Morton Morris, and the most visible part of Hazlitt is run

* Patrick's father, Francis Matthiesen, arrived in London in 1936 as a refugee from Berlin. He came via Zurich and opened a gallery behind the Connaught Hotel on the site of the Hamilton Gallery. This ran between 1936 and 1938, dealing in old masters, impressionists and drawings.

by James Holland-Hibbert, who sells Modern British art at the highest level.

A dealer who thrives on attention is Philip Mould, who brought art dealing to television with his long-running series *Fake or Fortune*, amusingly described by A. A. Gill as 'Gay Top Gear'. Mould has ploughed his own furrow, dealing in historical portraits, most recently from a gallery on Pall Mall. His brother, Anthony Mould, operates more privately, at the opposite end of the spectrum from Richard Green. If the latter follows the market, Anthony Mould offers his own sometimes quirky taste in eighteenth-century English pictures, with a special interest in Stubbs. He nurtures a small group of sophisticated clients, like Jonathan Ruffer and Lord Sumption, who have been happy to follow his taste.

One of the dealers who put Modern British art in the big league is Ivor Braka. Never feeling the need for a West End gallery, he has carved out his own niche and preferred to work from home. One of the most distinctive and admired dealers in London, Braka is sometimes mistaken for a pop star with his rock'n'roll appearance, big hair, charisma and easy charm. Educated at Oxford, trained at Sotheby's and Crane Kalman Gallery, he was brought up in Cheshire, where his father collected Lowry. Mentored by Andras Kalman and Anthony d'Offay, whose curatorship he admired, Braka started dealing mostly in British artists. These include Wyndham Lewis, Ben Nicholson, Frank Auerbach and Francis Bacon, alongside European minimalists like Piet Mondrian, stimulated by a visit to Count Panza's collection at Varese. Braka stresses the element of chance in dealing and has been adept at seizing opportunities, meeting many of his best clients at dinner. Attracted by the idea of someone who operated outside the gallery system and could advise them across the art market, his earliest clients included the insurance tycoons Ian Posgate and Robert Hiscox. After dinner with Hiscox, Braka would take him home, open a bottle of wine and, before the night was out, he had sold several hundred thousand pounds worth of art works. This became a pattern with many of his clients.

Francis Bacon has been perhaps the biggest theme of Braka's dealing. A very good working relationship with the Marlborough Gallery meant that he sold over forty works by the artist until they became too expensive. Braka believes that the art world changed in around 1995: those who bought before this point were collectors by and large, while afterwards art has been seen more as a brand and thus attracted a different audience. These new buyers are imbued with what Braka sees as 'the auction mentality', preferring to compete against other collectors and pay the buyer's premium rather than give a dealer a profit. This tendency contributes to the further demise of traditional dealers and galleries.

If you had walked down Bond Street and the adjoining Albemarle and Dover Streets in the 1970s, art galleries dominated the scene selling old masters and nineteenth-century paintings, interspersed with furniture and silver dealers. By the millennium, however, the landscape had changed, and fashion, already a strong component of Bond Street, had taken over. A generation of art dealers was retiring, and the new generation focused on contemporary art, which spread its tentacles wherever there was space big enough to accommodate it. These new galleries extended from the mewses and warehouses behind Bond Street all the way to Hoxton in one direction and the South Bank in the other.

Chapter 27

The Rise of Contemporary Art

'Contemporary art is the new Old Masters.'[1]

Simon de Pury

The rise in contemporary art represents the greatest shift in taste during the period covered by this book. In the course of twenty years, London went from the world centre of old master paintings sales to a major contemporary art city. Foundations had been laid at Whitechapel Gallery and the ICA, as well as at commercial galleries like the Marlborough, Kasmin and Robert Fraser. In 1982, the Tate Patrons of New Art was founded, uniting the new high society with contemporary art, and two years later the Turner Prize was instituted. The cosy world of Cork Street was expanded by Leslie Waddington, and beyond it a group of brilliant, often eccentric, dealers, including Nicholas Logsdail, Karsten Schubert and, above all, Anthony d'Offay, promoted both native and international talents. Charles Saatchi nourished the YBAs, a younger generation of British artists that coalesced around Damien Hirst. The foundation of art fairs and Sotheby's and Christie's forays into the contemporary market are important, while the period was crowned by the appearance of Tate Modern in 2000, and Larry Gagosian opening a gallery in London in the same year. Alongside this activity, a new generation of collectors emerged. Brought up on pop music, contemporary art spoke to them in a way that old masters did not.

Edmondo di Robilant, an Italian dealer in London, once observed that, 'The English, they don't collect except this strange thing they call Mod Brit.'[2] The exact point at which native contemporary art became 'Mod Brit' and flourished into a secondary market has never been defined, but the dealer who marches on both sides of the frontier and lived the transition most fully was Leslie Waddington (1934–2015). The engine of the London contemporary market, Waddington was, as the Tate director Nick Serota put it, 'on the front line for more than fifty years, a campaign veteran without equal in the profession'.[3] King of Cork Street, Waddington had, at his height, five galleries there and employed sixty people, although his reputation has been subsequently overshadowed by comparisons with Anthony d'Offay. An open and self-proclaimed *homme du peuple*, Waddington wanted to demystify art, claiming that, 'I think art galleries should be not so much like clubs as like supermarkets.'[4] If d'Offay was exclusive and secretive, Waddington was contemporary art's equivalent of Richard Green, with unfussy, simple presentation and a large, visible stock from which, in his time, he sold more contemporary art than any other dealer in London.

To the artist Michael Craig-Martin, Waddington was a brilliant dealer who could be stubborn, mercurial, opinionated and outspoken, but was always warm-hearted. A difficult man to work for, he nevertheless inspired great loyalty and affection.[5] In his own words, 'I am very difficult. I'm incredibly difficult. But I'm not charmless, not at all.'[6] Leslie, born in Dublin in 1934, studied history of art in Paris and contemplated becoming an academic. In his heart, he remained a Francophile, and one dealer suspected the Irish intellectual's disdain for England. For Nick Serota, 'Leslie's spiritual and intellectual roots lie in the literature rather than the art of Ireland, especially in the writings of James Joyce and in the modernist tradition as it developed in Paris between the wars.'[7]

In 1927, Leslie's father Victor opened the Waddington Gallery in Dublin, moving to Cork Street in 1957, and was best known for selling Jack Yeats. Leslie opened his own Cork Street space in 1966 with Alexander Bernstein as business partner. Among his first clients were the major collectors Ted Power and Alistair McAlpine, who donated his contemporary sculpture collection to the Tate in 1970. Waddington staged early shows of sculpture by William Turnbull, William Tucker and David Annesley, alongside works by Patrick Heron, Roger Hilton and Terry Frost.

Today, Waddington Gallery is best remembered for dealing with Baselitz, Flanagan, Craig-Martin, Caulfield, Blake and Waddington's older gods: Leger, Matisse, Picasso and, above all, Dubuffet, who was part of the gallery's DNA. If, individually, the exhibitions did not have the seminal quality of d'Offay's, as Thomas Lighton, one of Waddington's former employees put it:

> For me it was the quantity, quality and breadth of the exhibitions that was extraordinary; in the late eighties there would quite often be two or three exhibitions running concurrently. I suspect some of the most groundbreaking exhibitions were the American Colour field painters in the sixties/seventies, New Generation sculptors in the sixties, German neo-expressionists in the early eighties, but you have to remember that alongside these were heavyweight stock exhibitions – Picasso, Matisse, Leger, Dubuffet, and Picabia amongst others. Some of the ones I liked best were mixed stock shows – there was a series called 'Groups' that ran from about the mid-seventies to mid-eighties which would include a wonderful diversity of works.[8]

Perhaps the most interesting of these was held in 1984, and featured Carl Andre, Dan Flavin, Donald Judd and Sol LeWitt.

Colleagues were slightly sceptical about Waddington's American forays. As one put it, 'Leslie tried to be hip with American artists like Schnabel, but despite being friendly with John Houston, he didn't understand America.'[9] Attracted by the Tooth archive, with its list of paintings and clients, he briefly merged with the gallery where Peter Cochrane had first introduced American art to London, but Waddington-Tooth lasted only two years. During the 1980s, Waddington promoted Agnes Martin and Milton Avery, handling the latter's estate. By working with Arne Glimcher at Pace, he was able to attract Rauschenberg, amongst others. His greatest American coup, however, was when Julian Schnabel came to London to decide whether to show with d'Offay or Waddington. Although Anthony gave him a Beuys drawing, it was Leslie who won the battle.

Waddington was rarely an innovator and preferred mid-career artists: 'I take on new artists all the time, but they are new only to me. They are artists who are already partly established, often artists who are showing at Pace or Mary Boone in New York, or who are in the Whitney Biennial. I don't take on artists straight out of art school.'[10] More startlingly, he asked himself: 'Do I like *most* of what we show? Of course I don't like *most* of what we show. I like *some* of what we show. It's all very good art, but no one who thinks could possibly like *most* of what we show.'[11] Influenced by the Royal Academy's 1981 exhibition *A New Spirit in Painting*, Waddington appeared to be more comfortable with painting than any other medium.* To Larry Gagosian, for example, he was specifically 'a picture salesman', which some would argue with.[12]

* He was much assisted by Hester van Royen, who worked for the gallery in the early 1980s.

With the largest stock of modern art in London, a huge inventory of between five hundred and a thousand items, two thirds of Waddington's turnover came from the secondary market. During the 1980s, 80 per cent of his sales were overseas, especially to the USA but with a growing share to the Japanese. The latter was a volatile market, and one colleague remembers that in 1985 less than 5 per cent of the gallery's turnover was from Japanese sales, in 1990 it was up to a third, while the following year it was virtually nothing.[13] At his zenith around 1989, Waddington was the biggest contemporary dealer in London, some even believed in the world, with a turnover of almost £75 million. That year he sold a vast amount to Bo Alveryd, the Swede who swept through London buying art in bulk on behalf of a group of investors. Flush with cash, Waddington went on a buying spree at auction and elsewhere, backed by a substantial syndicated bank loan facility. However, the turn of the market meant that, by the end of 1990, Leslie was struggling and owed the banks £30 million.

Forced to sell paintings for half of what he had paid for them, Waddington found it harder to replenish stock and the gallery lost momentum. With his reduced sales he sometimes spent the afternoons playing chess with Alistair McAlpine or backgammon with Charles Saatchi. Although the gallery never recovered its 1980s level and the next decade would belong to d'Offay, Waddington survived in part thanks to his stable of artists: Flanagan, Heron, Blake, Hitchens, Frink and Caulfield. Michael Craig-Martin drew Waddington's attention to his remarkable group of students at Goldsmith's, and he gave Ian Davenport his first solo show in 1990, and exhibited Fiona Rae the following year. During the 1990s, the market recovered, and like d'Offay, Waddington benefited from the new rich Americans and continentals who came into London.

Despite Waddington and d'Offay's dominance over the market, young artists' favourite destination was Nicholas Logsdail's Lisson Gallery. As Craig-Martin explained, 'Lisson was the goal of a generation of British artists; Anthony d'Offay was too grand, Leslie Waddington was too stuck in the pop generation and Colour field, and too rooted in the 1960s, so Lisson was the one.'[14] To Karsten Schubert, who trained there, the gallery was 'where British art of the 1980s happened'. There was an intellectual rigour about Lisson, which was known for championing minimalism and conceptual art on one hand and British sculpture on the other. Logsdail promoted Tony Cragg, Richard Deakin, Anish Kapoor, Bill Woodrow, Edward Allington, Richard Wentworth and Richard Long. As he reflected, 'In retrospect, I have come to believe that you make your luck through your interests which resulted, in this case, in Lisson's inception.'[15]

Logsdail was first inspired by modern art thanks to his uncle Roald Dahl's art collection, which included Francis Bacon. Initially, he wanted to be an artist and enrolled at the Slade, where his interests shifted. After a night of drinking, he missed the last train home and slept on a bench by Lisson Grove. Beside the bench was a four-storey house with a sign that read 'For Sale, No Reasonable Offer Refused'.[16] He bought the property for £2,000 and opened his gallery in 1967 with a show of five young artists, including Keith Milow and Derek Jarman. Just off Edgware Road, the gallery was outside the usual topography of the 1960s art trade.

Logsdail's interest in American minimalism was somewhat pioneering in London. As he explained:

> I began going to New York myself at the end of the 1960s and in some ways never looked back. I understood the importance of what Marshall McLuhan called 'The Global Village' and for

> the Western art world, facilitated by inexpensive travel, the internationalist ideal found its home in New York ... When I went to New York for the first time and I told my friend, Barbara Reis, what I saw, she said 'Well, did you go and see the artists?' and I said, 'I didn't have their phone numbers', to which she responded, 'You dumb klutz, they are in the phone book'. I felt so stupid. But actually, I'm not sure at that time I had the courage to call them up to tell them I was interested, and that I had a small gallery in London. When I next returned to New York, she was absolutely right. The artists were delighted to hear from me. Few from London had expressed interest in them. The London art world, as it existed, did not pay much notice to the artists we were showing at the Lisson Gallery for at least the first decade. They didn't yet know of Sol LeWitt, Donald Judd, Carl Andre, Robert Ryman, Lawrence Weiner – or even Richard Long, outside of my generation.[17]

After a trip to New York in 1968, Logsdail staged a joint show of Donald Judd and Sol LeWitt, the first of many UK debuts for major American minimal and conceptual artists. He found a hungry client in the voracious Italian collector Count Panza. Logsdail took some interesting risks, including the *Wall Show* (1970), a group exhibition in which twenty artists were given a blank wall each, which resulted in seminal works like Lawrence Weiner's *A Removal From The Lathing Or Support Wall or Wallboard From A Wall* (1968), now in the collection of the Museum of Modern Art (MoMA).

In the 1980s, Logsdail moved into British sculpture, showing Richard Deacon, Shirazeh Houshiary and Julian Opie. Two of his sculptors, Tony Cragg and Anish Kapoor, won the Turner Prize in 1988 and 1991, respectively. In 1991, the gallery moved to nearby Bell Street, allowing Logsdail to exist 'in splendid isolation from Mayfair and to create our own art world'.[18]

Lisson has a unique and important place as a fascinating gallery of exquisite taste. Although the YBAs were knocking at his door, the cerebral Logsdail did not engage with them in any major way.*

Harder to assess today is Nigel Greenwood, who Logsdail described as 'a kindred spirit and we were almost firm friends, we were maybe barking up the same tree and pissing on both sides of it at the same time. But we too had different tastes, for example, he was showing Gilbert & George and Marc Camille Chaimowicz, while Lisson was showing Richard Long and Sol LeWitt.'[19] In 1971, Greenwood established his gallery at 41 Sloane Gardens, after studying at Oxford and the Courtauld. He was by all accounts 'a charming rather shy man, too decent for the art world, who sometimes found it difficult to pay his bills'.[20] As Nick Serota wrote:

> Greenwood's tastes were broad, and he showed a bewildering, even erratic, range of artists, all of whom, like himself, were marked by their independent-mindedness. He was one of the few British gallerists to look as much to Europe as to America in the 1970s, as enthusiastic for unknown artists as for those with international reputations ... He may have been better at discovering artists than keeping them, but if not for him several major careers would not have been launched, others would not have been sustained through lean years.[21]

Perhaps Greenwood's achievement is best summed up by the fact that Gilbert & George, Keith Milow, David Tremlett, Rita Donagh, Christopher Le Brun and Dhruva Mistry all first exhibited at his gallery. In 1985, he moved to New Burlington

* In 1993, the gallery staged a show entitled *A Wonderful Life*, which included Damien Hirst and Martin Creed alongside Jason Martin and Simon Patterson.

Street and the same year he was given the accolade of selecting the Hayward Gallery Annual Exhibition, the only dealer to be so invited. The gallery became a victim of the 1990s downturn, and Greenwood eventually sold his stock through Sotheby's.

It is impossible to discuss the London contemporary art market without mentioning Art Basel, the most important of the growing number of international art fairs, most of which operate outside the timeframe of this book. Basel opened in 1970 and was initially patronised by German dealers, although by the end of the decade most of the main London dealers – Marlborough, Waddington, d'Offay, Mayor, Lisson, Annely Juda and Greenwood – had stands there. The Basel dealer Ernst Beyeler was the prime mover, and the fair cleverly promoted newer artists, validated by the presence of works by Picasso and Henry Moore.

Founded in 1989, the London Art Fair was staged in the Business Design Centre in Islington. It was never to have the glamour of *Frieze*, which began life in 1991 as a magazine, founded by Amanda Sharp, Matthew Slotover and Tom Gidley. Oxford graduates with no experience of the art market, they put Damien Hirst's *Butterfly* on the first cover and published an interview with the artist. In 2003, the magazine turned into an art fair of the same name, held in a tent in Regent's Park designed by the architect David Adjaye.

In 1985, artists' disillusion with traditional galleries gave rise to the 'Cork Street Attack', as the 'Grey Organisation' splashed paint over every gallery window on the street. The organisation consisted of a group of young artists intent on cocking a snook at the establishment and received a banning order from central London. In fact, hip galleries were increasingly opening in districts closer to where most artists had studios. The move to open spaces in the East End began as early as the 1970s and gathered pace in the 1990s.* Although the relocation began with small start-ups, West End galleries also sensed an opportunity in the East. White Cube is the most famous example, and in 2000 a satellite of Jay Jopling's Duke Street gallery opened at 48 Hoxton Square.† However, in the new millennium, 'The big East End galleries rather ran out of steam. Galleries went there but found that American and European collectors didn't venture out to Hackney. When Larry Gagosian opened his gallery at King's Cross, he found there was resistance to getting collectors there.'[22]

As a primary market, contemporary art was initially the dealers' preserve and auction houses stepped falteringly onto the scene. The first major contemporary art auction was the legendary sale of Robert and Ethel Scull at Sotheby's New York in 1973. It was controversial – interestingly more with artists than dealers – and very successful. Christie's first European contemporary sale was organised by their local representative, Jörg Bertz, in Düsseldorf on 14 November 1973, and followed up in London the next year. Sotheby's began selling contemporary art around the same time, including it in the afternoon session of their impressionist and modern auctions. Initially, none of this amounted to much, but that was to change dramatically when, in 1988, Jasper Johns's *False Start* (1959) made $17,050,000 at auction, signalling a new price level for a living artist. By the end of the millennium, contemporary auctions were overtaking impressionism as the flagship sales of Sotheby's and Christie's.

The growing interest in contemporary art provides the background to James Roundell's observation that, 'Traditionally,

* Robin Klassnik opening Matt's Gallery in Hackney in the 1970s is sometimes seen as the pioneer.

† Jay Jopling would close this gallery but open in Bermondsey in 2011.

Sotheby's and Christie's fought each other for about 40 per cent of the market. Then in the early 1990s, they realised that the growth would come from the other 60 per cent, the part held by dealers.'[23] Their initial solution was to acquire contemporary art dealerships. In 1996, Sotheby's bought the André Emmerich Gallery in New York as a vehicle for dealing with artists and their estates.* Christie's did not follow this move until after the French tycoon François Pinault had taken control of the auction house in 1998. Four years later, Pinault went to view the Rachel Whiteread exhibition at the new Haunch of Venison Gallery, set up by Harry Blain and Graham Southern. Pinault bought *100 Spaces*, previously owned by Charles Saatchi, and became a regular customer of the gallery. This began a relationship which led to his acquisition of Haunch of Venison as a vehicle for undertaking private sales in contemporary and modern art for Christie's. Such steps allowed Sotheby's and Christie's to move into the primary market and private sales.

The founding of Tate Modern in 2000 was by far the most important cultural event of the era, and one that represents London's coming of age as a major force in contemporary art. As Larry Gagosian put it, 'Tate Modern signaled London to be taken seriously – artists were keen to engage with it – a game changer for London.'[24] After Big Bang, the city was attracting billionaires, many of whom gravitated towards the museum and to Nick Serota's Patrons of New Art. Over the decade following the founding of Tate Modern, international mega-dealers arrived in London, starting with Gagosian in 2000, seeing the city as a gateway to Europe.

When Gagosian opened on Heddon Street, he was already the world's most successful art dealer. His hero has always been the pre-war British dealer Joseph Duveen. As Gagosian told the

* In 1990, Sotheby's had already teamed up with Bill Acquavella to form Acquavella Modern Art, a subsidiary of Sotheby's Holding Company.

author, 'Americans are comfortable in London – London was my port of entry to the European market. The auction houses added to the stature of the city.'[25] His first forays into London had been via Molly Dent-Brocklehurst, a well-connected agent operating from a small office in Berkeley Square. Marc Glimcher of Pace Gallery commented that when 'Larry opened in London in 2000 – he didn't even know how brilliant a move that was. He was stepping into the first wave of Russians coming to London. He had kind of stepped right into it.'[26] The d'Offay Gallery closed in 2001 and, as Stefania Bortolami later observed, 'Larry got lucky. He was expanding just as d'Offay closed.'[27] That said, Gagosian's arrival may itself have been a contributing factor to d'Offay's closure.

Hauser and Wirth opened in London in 2003, and Pace and David Zwirner in 2012. For the first time, the most important dealerships (with a few exceptions, like White Cube) had their headquarters elsewhere. For most of the local dealers, however, their arrival was a positive development. Nicholas Logsdail thought the big players brought enhanced energy and range to London, as well as new artists: 'I welcomed the endorsement of London and of contemporary art they brought with them. In many ways their expansionism spurred us on to open spaces in Milan, New York, Shanghai, Beijing and now, Los Angeles.'[28] Seen in Hollywood terms, the arrival of such international art brands was the final victory of studio over independent. London was connected by these major dealerships of the twenty-first century to galleries all over the world, sharing with them the museum-like display, sales teams and even a futures market of as-yet-uncreated works of art.

Chapter 28

Anthony d'Offay

'A d'Offay show was always an event. Anthony behaved as though he was supporting a cause.'[1]

Nick Serota

Anthony d'Offay, described as the sphinx of the art world, has always been an enigma. One of the most discussed dealers in this book, he is loathed by some, and admired by most. D'Offay's monkish, messianic – almost ruthless – character has driven his sense of mission. When in pursuit, his ability to make friends was equal to his ability to lose them afterwards, hence his moniker 'Prince of Darkness'. Behind this enigmatic façade lay the most interesting intelligence in the London art market. For d'Offay, art is a sacred trust, and he retained a childhood conviction that it can transform lives, staging exhibitions with an almost sacramental seriousness. When Larry Gagosian was asked which London dealer he most admired, he unhesitatingly replied, 'Anthony d'Offay: his style, his artists, all those little galleries in the same area, his minimalist approach to dealing.'[2] D'Offay's singularity is underlined by his extraordinary trajectory, which reads like a nineteenth-century French novel. From bookish provincial teenager, he rose to become king of the London contemporary scene, an ascendency that ended with his sudden unexplained closure in 2001.

Over his career, no dealer caught the zeitgeist better than d'Offay, catering to the aspirations of the embryonic Tate Modern and contributing most to its collection. He was always in motion, believing that 'You have to reinvent the gallery every seven years'. Illustrating this, he moved from Eric Gill and David Jones, via Beardsley and Japanese prints, to Modern British art, vorticism and Bloomsbury, to his final flowering in the mid-1980s as the encyclopaedia of international contemporary art. No London dealer made a broader transition or showed such a variety and quality of international contemporary art. The contemporary art world's question was always what will Anthony do next? Sadie Coles, who once worked at d'Offay Gallery, said that his great talent was to surprise. His exhibitions had a seminal importance and the two turning points in his career are reflected through his *Abstract Art in Britain* exhibition in 1969, and his Joseph Beuys exhibition in 1980. Both shows signalled his direction of travel for the decade that followed.

Anthony d'Offay was born in Leicester in 1941, the son of a French surgeon father and an English antique dealer mother. His first exposure to art was when she would drop him off at the local museum (then arranged by the émigré curator Hans Hess) while she went to run errands. While at Edinburgh University, d'Offay discovered the Scottish National Gallery's collection of old master paintings, calling it 'the experience that defined my life'.* His first interest, however, was in the art and literature of the 1890s, and he began buying and selling work by Aubrey Beardsley. Richard Demarco, who ran a gallery in Edinburgh at the time, recalls meeting d'Offay in the early 1960s: 'This extraordinarily serious young man came into my gallery,

* Years later, bringing Damien Hirst's pickled sheep and Robert Mapplethorpe's sado-masochistic photographs to Scotland, he hoped they would 'make a great difference to young people, their appreciation of themselves and their era. It will inform what they do with their lives and how they see the world.'

endeavouring to interest me in a print. It seemed to me remarkable that he was so absolutely committed to it.'[3]

D'Offay arrived in London in 1964 and was initially more of a book dealer than an art dealer, handling the library of Canon John Gray, which focused on the aesthetic movement. He applied to work in Sotheby's book department but was told: 'you are a dealer'. In 1965, he opened a tiny gallery in Vigo Street, off Regent Street, selling rare books, drawings and Japanese prints, with an emphasis on the 1890s. His first major catalogue, 'Art and Literature 1870–1920' (1965), included works by Firbank, Huysman, Cocteau and Proust. Over time, d'Offay moved closer towards the visual arts, with exhibitions such as *Dream and Fantasy in English Painting 1830–1910* (1967), in which he revived interest in John Martin and Richard Dadd.

His gallery in Dering Street launched in 1969, with the pioneering exhibition *Abstract Art in Britain*, which included works by David Bomberg, Jacob Epstein, Frederick Etchells, Henri Gaudier-Brzeska, William Roberts, Helen Saunders and, above all, Percy Wyndham Lewis. Every piece was meticulously chosen, and the catalogue was designed by Gordon House, who made Marlborough's catalogues. D'Offay began to specialise in vorticism, Camden Town and Bloomsbury, which he elevated in line with the efforts of the Tate curator Richard Morphet. One of his best-remembered exhibitions was of Gwen John (1976), after one of his staff, Stephanie Maison, managed to secure her estate from her nephew, Edwin John. Such was the quality of his exhibitions – and his persuasiveness – that d'Offay was able to borrow the best of an artist's work from national collections, adding depth to the work for sale. In tune with his dictum of changing focus every seven years, during the 1970s d'Offay started to show contemporary art: Lucian Freud, Michael Andrews, Eduardo Paolozzi, Frank Auerbach and William Coldstream. Like Waddington, he was much influenced by the RA's 1981 *New Spirit in Painting*, which included contemporary works from Germany, Italy and the USA. Organised by Nick Serota, the RA Exhibitions Secretary Norman Rosenthal and the curator Christos Joachimides, the show became the intellectual backbone of what d'Offay would sell over the next decade.

D'Offay's marriage to Anne Seymour in 1977 was to signal one of his most decisive changes of direction. In his own words:

> Anne was one of the curators of the modern collection at the Tate, and she made it her ambition to change me into a modern person. So there were three of us – myself, my wife Anne, and Marie-Louise Laband, who joined in 1976 and ran the operation faultlessly for the next four decades. Our first exhibition when we opened in 1980 was Joseph Beuys. The new gallery had eight rooms, some of them really large. We had the idea that if we could have a big gallery, like the New York Soho galleries of the time, we could invite the world's greatest living artists to show here, and that this new programme would help to make London a destination for the international art world. We imagined that people would come to London to see those shows and it would have a great impact, and I think that really worked. Beuys's *Stripes from the House of the Shaman* was the opening exhibition, and this was Anne's idea.[4]

Anne Seymour's arrival transformed the gallery. An experienced curator, she spoke German, having lived in the country as a child, and opened the way to a more international market. D'Offay's inaugural Beuys exhibition was the platform that took him into international contemporary art. As d'Offay claimed at the show's opening, 'We declare ourselves to be in the world rather than in London.' Richard Demarco, who had introduced Beuys to Britain, explained how the exhibition came about: 'Joseph needed £2.5m to fund his planting of 7,000 oak trees for a monumental piece so he asked me if I would not mind Anthony d'Offay representing him from London. We went down to Dering Street and took tea.'[5] Beuys drew other artists to d'Offay, including Anselm Kiefer in 1982, then considered by many to be the most interesting European artist. Kiefer confessed to d'Offay that 'the most important thing in my life was spending time with Beuys'.[6] Beuys would later be central to d'Offay's Artist Rooms, in which he would be represented with 136 works. Exhibiting Georg Baselitz was another coup for d'Offay, who was aided in his German endeavours by the gallerist Helen van der Meij.

Richard Calvocoressi, then a curator at the Tate Gallery, recalls:

> For me, the real game-changer was d'Offay, who became international in 1980 when he (with ex-Tate wife Anne) opened his second, larger gallery in Dering Street, showing Beuys, Kiefer, Baselitz, the Italians, Twombly, Warhol, Johns, etc. I always felt his programme was more focused than Waddington's. He was also keen to get his artists into museum collections and was prepared to wait if funding was a problem – unlike some other dealers. I experienced this at the Tate with Beuys, when Anthony took me to visit him in Düsseldorf (in 1984, I think) and persuaded Beuys to donate a handful of small objects to make up a third vitrine in addition to the two that the Tate was buying (from d'Offay). I would say Anthony's patience and long-term view benefitted public collections.[7]

As Nick Serota pointed out, an exhibition at d'Offay was always an event, and somebody compared 'Waddington's supermarket against d'Offay's delicatessen'.[8] D'Offay's normal method was to buy a work and build an exhibition around it. Usually held twice a year, the shows mixed loans from the artist, their family or museums, alongside the items for sale. To persuade artists, one former colleague suggested that, 'Anthony would buy one important work by an artist at auction for a large sum which would gain the attention of the artist. He would then propose to do an exhibition around the work, requesting additional items from the artist.'[9] All contemporary dealers need to build a good relationship with their artists, and d'Offay knew better than most how to seduce them through the quality of his exhibitions and his focus. Acting as a curator searching for perfect pieces and crafting his exhibitions with museum-style cataloguing, artists were impressed by his high seriousness and his tendency to help get their work into museum collections. D'Offay went to great trouble to find thoughtful gifts for his artists. When d'Offay was trying to woo Beuys, Nick Serota recalls him asking his German dealer, Heiner Bastian, what kind of hats or shoes Mr. Beuys would like.[10] Despite his persuasiveness, d'Offay was never a clubbable, sociable dealer like Leslie Waddington. Cerebral and remote, he is more often described as an *éminence grise*.

Some said that d'Offay Gallery resembled a sect, and although the morning meetings 'were not prayer meetings they were conducted in a very controlled way. Anthony's absorption amounted to a compulsion, almost obsession. His staff are either with him 150 per cent and so take calls at all times of day and night or cast out.'[11] Over the years, d'Offay employed an extraordinary number of talented people: Robin Vousden, Sadie Coles, Matthew Marks, Simon Lee, Sarah Lucas, James Cohan, Lorcan O'Neil and many others.

During the 1980s, he became increasingly focused on the USA and went there once or twice a month 'to teach myself to think like an American, to make great friends with the museums'.[12] In fact, it was the American artists and dealers who would be as important to him. When d'Offay told Anselm Kiefer that he was going to show Jeff Koons, Kiefer said, 'Well, I hate his work, but I think it's a very good idea for the gallery.'[13] D'Offay formed very effective relationships with two New York galleries. The first was Xavier Fourcade, through whom he first showed

Willem de Kooning. More important in the long run was Arne Glimcher at Pace Gallery, who gave him access to American artists and clients. A large part of d'Offay's success came from selling American art to Germany and German art to America. As one admirer put it, 'he saw London as a rocky outpost'.

In the mid-1980s, d'Offay was enthusiastic to stage a Warhol show in London. Introduced to Warhol through Beuys, he cultivated the relationship, regularly visiting the artist in New York. Asking Warhol, 'What subject would you like to try next?' the artist replied, 'Oh Anthony, you choose.' D'Offay suggested self-portraits, to which Warhol responded, 'Great! Maybe we can do something with camouflage?'[14] At first dubious, d'Offay was eventually delighted with what became known as the 'Fright Wig Self-Portrait paintings'. The exhibition opened in the summer of 1986, and the pictures are seen as Warhol's last great works before he died the following year. For him and other artists, part of d'Offay's appeal was his willingness to invest in their projects without restriction.

D'Offay exhibitions through the 1990s were beautifully curated compendiums of American, German and British art. Asked about the artists he was showing that decade, d'Offay rolled off a list:

> From Europe: Joseph Beuys, Gerhard Richter, Anselm Kiefer, Sigmar Polke, Georg Baselitz and Jannis Kounellis. From America: Willem de Kooning, Cy Twombly, Jasper Johns, Ellsworth Kelly, Andy Warhol, Roy Lichtenstein, Vija Celmins, Carl Andre, Lawrence Weiner, Bill Viola and Jeff Koons. From the UK: Richard Hamilton, Howard Hodgkin, Richard Long, Gilbert & George and Ron Mueck. We tried to think of who were the world's most interesting artists.[15]

D'Offay's relationships with British artists were famously up and down. His fallings-out are legends of the art world: Gilbert

& George (who he had taken over from Nigel Greenwood) stomped off to join Jay Jopling. Grayson Perry's parting gesture was a pair of large black pottery penises entitled *Portrait of Anthony d'Offay*. Lucian Freud, notoriously difficult at the best of times, was another case. D'Offay began to share the artist with James Kirkman in 1970, and at first all was well, with d'Offay describing the arrangement as, 'A wonderful experience. Such an extraordinary person, a great artist and friend. He was the best man at our wedding.'[16] As the dealer recalled, 'In 1988 we showed an Australian performance artist called Leigh Bowery. Freud saw him, became fascinated by him, and started to make unusually large nude paintings of him.'[17] Although Freud was inspired by d'Offay exhibitions, his own shows were, however, rarely rewarding for the dealer, because he was such a slow worker, producing a picture every two months, which was usually already sold. In addition, Freud would often need money the instant a picture was finished to feed his gambling habit. The falling-out came when d'Offay truncated one of Freud's shows to make way for a Baselitz exhibition.

Following this, Freud contacted the New York dealer Bill Acquavella. Knowing the artist was tricky, Acquavella was sceptical until he went to his studio and saw a large painting of the naked Leigh Bowery, the power of which knocked him out. He offered to buy everything Freud made with no contract on either side, and he did exactly that over the next two years. Within a year, Acquavella found himself paying off huge gambling debts for the artist. However, the collaboration was an international breakthrough for Freud and changed his career. Describing Freud as 'the greatest living painter', the dealer found buyers for his very large canvases at unprecedented prices. There is no doubt that Acquavella pushed the artist's prices far beyond what Kirkman and d'Offay had been able to achieve.

Freud was not the only artist influenced by d'Offay's Leigh Bowery exhibition. The Australian performance artist sometimes worked behind a glass screen from which he couldn't see out, but people could look in. His performance at the gallery fascinated Damien Hirst, who worked there as a technician, painting walls and observing how the art world operated. Although d'Offay never handled Hirst's work, he would acquire some pieces for the Artist Rooms project. D'Offay's association with the YBAs would be relatively slight, and usually at the encouragement of Sadie Coles. Michael Craig-Martin believes he was puzzled by them. Despite this, Marie-Louise Laband, who worked with d'Offay, recalls that 'a lot of artists approached us', and the gallery exhibited Gary Hume, Rachel Whiteread (taken from Karsten Schubert, who never forgave Anthony) and Ron Mueck. Included in Charles Saatchi's *Sensation* exhibition at the Royal Academy in 1997, Mueck's sculpture *Dead Dad* became one of the stars of the show, and d'Offay gave the artist his first solo exhibition the following year.

The most pored-over event in d'Offay's history is the sudden closure of his gallery in 2001. He had just acquired a big space at Haunch of Venison Gallery and employed Graham Southern from Christie's. Exactly what happened is unclear. Some think that he lost his nerve, as a number of artists went to other galleries. Others say he realised that his next move would cost a fortune. At the same time, the mega-dealers were sizing up London and d'Offay may have concluded that 'the game was up', as one of his artists put it. Gagosian had arrived and Hauser & Wirth were planning to come, jeopardising d'Offay's virtual monopoly.

It seems that d'Offay had a choice: to gear up and go head to head with these well-financed international dealers, or to retire, become a myth and a philanthropist. Whatever the reasons, d'Offay chose the latter course, making a deal with Edinburgh National Gallery of Modern Art and Tate to realise his vision of the Artist Rooms project. D'Offay's subsequent story as begetter and curator of Artist Rooms is beyond the scope of this book.

His developing friendship with Richard Calvocoressi at the Scottish Gallery of Modern Art (in the city that nurtured him intellectually as a student) gave him a vehicle to demonstrate his belief that art was better seen in rooms, and he conceived the idea of the series in around 2003. In 2008, his dream of being a curator was realised when he transferred (at a fraction of their value) 725 works by twenty-three artists to the joint ownership of the National Galleries of Scotland and the Tate, filling many gaps in the collections of the two museums.

In d'Offay's own words, 'We had run the gallery for 21 years and felt we had completed what we had set out to do. I did not want the gallery to become tired or repetitive and the only way to close was to do so suddenly at the beginning of the new season, in September 2001.'[18] On a personal level too, he announced 'I feel complete'. All should have been serene for this next phase of his life, but allegations of sexual harassment from three women he worked with, including the artist Jade Montserrat, damaged his reputation, and his name was for a time removed from the Artist Rooms. The events create a tragic coda to the most intelligent and interesting dealer in the second half of this book.

Chapter 29

The YBAs

'When I saw the shark for the first time, I thought this is the most impactful encounter I have ever had with a work of art.'[1]

Larry Gagosian

The group of predominantly Goldsmiths students who came to be known firstly as the 'Brit Pack' and then as the YBAs (Young British Artists) burst onto the scene at *Freeze* (not to be confused with the later Frieze), a self-organised exhibition held in 1988. They did not, however, gain wider international recognition until their work was showcased in *Sensation*, the 1997 Royal Academy exhibition. A small number of YBAs, especially their ringmaster Damien Hirst, became part of the canon of what every contemporary art collector and museum wanted. Never in the history of British art has an artist who initially set out to prove that he did not need the art establishment so comprehensively seduced and disrupted it. Within two decades, Hirst's takeover held Charles Saatchi, the Royal Academy, Tate Modern, White Cube and, finally, Sotheby's in thrall to his energetic talents.

No British collector during the late 1980s and early 1990s was more powerful than Charles Saatchi, or more stimulating to a younger generation of British artists. Saatchi was not a collector in the traditional sense – he bought art to show it. His exhibition *kunsthalle* at Boundary Road in St John's Wood became the showcase for new ideas and artists. London had seen nothing like it before. A large, neutral space dedicated entirely to contemporary art, the Saatchi Gallery became a focus of the London avant-garde before the opening of Tate Modern. As far as Larry Gagosian is concerned, 'Charles Saatchi's contribution cannot be overstated – he put London on the map. Boundary Road showed great collecting, extremely important in developing London and young artists. We became good friends, and he would come to New York and buy whole exhibitions.'[2] For his part, Saatchi once jokingly told an interviewer, 'Whenever I see Gagosian approaching, I hear the soundtrack of *Jaws*.'[3]

Charles and his brother Maurice founded their advertising agency Saatchi & Saatchi in 1970; by the end of the decade, it was the largest in Britain, and in 1979 it devised the campaign behind the Conservative Party's election to power, encapsulated in the billboard 'Labour isn't working'. As a child, Charles Saatchi had collected Superman comics and juke boxes. His first wife, Doris Lockhart, whom he married in 1973, came from Memphis, Tennessee, and together they started buying minimalism from the Lisson Gallery: Carl Andre, Donald Judd and Sol LeWitt. The focus was initially on America, and early on Saatchi established a pattern of making multiple purchases.

Saatchi opened Boundary Road in 1985 with a show of American art: Donald Judd, Brice Marden, Cy Twombly and Andy Warhol. The next year, this was followed with Carl Andre, Sol LeWitt, Robert Ryman, Frank Stella and Dan Flavin. The American artists he chose were already well known in New York, but were new and influential to the young British artists who saw these exhibitions. Damien Hirst recalled, 'When I saw the Saatchi Gallery, I thought, "Here's a place for that sort of scale". I remember walking into a Saatchi show for the first time and feeling, like, snow blind. I'd never really experienced anything like that.'[4] Collectors are usually credited for their choice of art works, but Charles Saatchi is one of only a handful who influenced the creative process of the next generation, nourished on his exhibitions. As Damien Hirst put it, 'I think Saatchi is one of the biggest, if not the biggest, reason for the YBAs.'[5]

Michael Craig-Martin, the brilliant, undidactic tutor at Goldsmiths, was the benign godfather to the YBAs.[6] In the mid-1980s, he was beginning to think that 'there was no possibility of radicalism. Everything looked like something you saw 40 or 50 years ago.'[7] In 1988, however, Craig-Martin realised he had encountered a talented and original new generation, with a self-appointed leader in Damien Hirst. As a student, Hirst worked part-time at the d'Offay Gallery, and as Craig-Martin noted, 'Damien was trying to figure out how [the art market] works.'[8] Nobody would understand and manage this market more effectively over the next twenty years.

The commercial art trade, especially Cork Street, seemed impossibly remote to most artists. Instead, the Goldsmiths students looked to the Lisson Gallery, hoping for a show there at the cutting edge of new art. The main problem Nicholas Logsdail faced at Lisson was that his gallery was small, and he already had a full roster. He told Craig-Martin's students 'perhaps in two, or three, years' time'.* As the critic Adrian Searle described, 'Apart from rare stars such as Julian Opie, who had been signed up by Lisson Gallery after graduating from Goldsmiths, art students faced years of penury, supporting themselves through odd jobs before they could hope for a show at a private gallery. Everyone was waiting in line, when do I get a show, do I have to wait till Howard Hodgkin dies?'[9]

Not content with this process, Damien Hirst had other

* Information from Michael Craig-Martin. To some extent, Logsdail did fulfill this promise. In 1993, Lisson staged a show entitled *A Wonderful Life*, which included Damien Hirst and Martin Creed alongside Jason Martin and Simon Patterson, among others of that YBA generation.

ideas. He and his fellow artists decided instead to bypass the gallery system and promote themselves. Hirst searched for a space similar to Boundary Road and found an empty ex-Port of London Authority building. In August 1988, they opened *Freeze*, an exhibition that would achieve legendary status. Sponsored by the London Docklands Development Corporation, the archetypal Thatcherite regeneration engine, the three-part exhibition was about visibility and gaining attention. Craig-Martin persuaded Norman Rosenthal and Nick Serota to see the show, and Rosenthal remembers being driven there by Hirst: 'I for one managed to see it courtesy of the persistent young Hirst, who came to collect me very early one morning in a rickety old car and drove me down to Docklands so that I might be back at the Royal Academy by 10:30am.'[10] With typical generosity, Hirst insisted on a detour to inspect *Ghost*, a large sculpture by Rachel Whiteread exhibited at the Chisenhale Gallery nearby. Rosenthal was impressed by all he saw, and the exhibition would blossom a decade later into *Sensation*.

Michael Craig-Martin is bemused that discussions of *Freeze* present it as a cleverly thought-out plan, rather than a combination of youthful bravado, naivety, good timing and, above all, interesting art.[11] Saatchi, still American-focused, was not immediately struck by what he saw, although he enjoyed what he called 'the hopeful swagger of it all'. Forced to sell a part of his collection to Larry Gagosian during his divorce in 1989, he decided to replenish his gallery from this new generation of British artists. Despite having circumnavigated the gallery system, *Freeze* intrigued some dealers. Jay Jopling, the dealer most associated with the group, was still finding his feet. This meant that it was Anthony d'Offay – who Craig-Martin described as a little baffled by the YBAs – and above all Karsten Schubert, a German dealer posing as a young fogey, who were the first onto the pitch.

Born in Berlin, Karsten Schubert (1961–2019) was probably the only London art dealer who had studied theology (at Humboldt University), before apprenticing with an art dealer in Cologne. When he came to London, he worked at the Lisson Gallery. Described as 'the picture of a certain kind of Englishman: small, solid, with unruly fair hair and gold-rimmed spectacles', he was an intellectual and a writer who was obsessed with Proust.[12] Schubert opened his first gallery in Charlotte Street in Fitzrovia in 1986, and thereafter opened and closed several times in tune with his changing fortunes. To Norman Rosenthal, 'Karsten was not a natural dealer and defined himself as a scholar.'[13] His star artist was Bridget Riley, whose success paid for his upcoming artists.

On Craig-Martin's advice, Schubert went to the Goldsmiths degree show and gave Gary Hume, Michael Landy and Ian Davenport their first West End exhibition. He also signed up Rachel Whiteread, whose *House* (1993) became as famous as Hirst's shark and Tracey Emin's bed. Although Whiteread won the Turner Prize on the back of the life-sized, site-specific installation, it was not a work that could be sold in any conventional way. Schubert made little or no money out of early Rachel Whiteread, and it was only when Anthony d'Offay began to make smaller casts of her pieces that she became commercially practical. Schubert was also the first dealer to buy Hirst's work, paying £500 each for the medicine cabinets and, according to the artist, later selling them to Jay Jopling. Hirst's relationship with Schubert was difficult:

> I was arguing with Karsten; I couldn't get into that gallery vibe … I had lunch with him and Helen Windsor, and I told him, 'I'm ready to come into the gallery'. He said, 'good. But I think what you need to do is go back to your studio and you need to make work for us for six months to a year, and when you've got something to show me, I'll come and have a look at it'. That was it. I was just like 'Fuck off. I'm never going there'.[14]

Schubert could be prickly and tactless with his artists, once famously sending a memo in which he mentioned the need to trim 'dead wood'. He lost Gary Hume and Rachel Whiteread, whose departure to d'Offay precipitated the closure of his gallery in 1996. Following this, Schubert concentrated on Ridinghouse, his publishing house and gallery formed the year before and best remembered for screening Jake and Dinos Chapman's pornographic film *Bring Me the Head of* (1995). Despite being cerebral and personally fastidious, Schubert had a baroque side that came out in his love of novelty. During a long illness, he lived at Claridge's Hotel, described in his last book, *Room 225–6: A Novel* (2015), even if he couldn't afford the laundry. Schubert never managed to benefit fully from his percipience, and Jay Jopling's easy-going sociability and optimism won over the YBAs.

Jopling's career has been charmed. Endowed with charisma and confidence in spades, he has only had to ask to receive. That said, nobody has worked and played harder in making White Cube the flagship British contemporary art gallery. Jopling's appearance has been compared to Robert Fraser's: 'Tall, dark suit, white shirt, black rim specs, with a slight air of inscrutability, both old Etonians.'[15] The most striking thing about him, however, is his ability to make friends with artists and enjoy socialising with them, all night if necessary. 'He's got a zest for life and he doesn't want to miss out on anything,' said the art historian Tim Marlow. It was this ability that brought him golden opportunities. Perhaps his greatest luck was to be in exactly the right time and place to facilitate Damien Hirst's fantastical concepts.

Like d'Offay, Jopling was a self-taught dealer who, in his case, was dropped off by his mother at the Tate as a child. Born in 1963, the son of Gail and Michael Jopling (a Yorkshire landowner and cabinet minister in Margaret Thatcher's government), Jay went to Eton and persuaded Bridget Riley to create a cover for the school magazine. The dealer who most inspired him was d'Offay, and he recalled that 'as a student you saw world class art at Anthony'.*[16] He was, however, equally interested by Charles Saatchi's gallery at Boundary Road. Convinced that he wanted to work in the art world, he studied art history at Edinburgh University, where he arranged a charity sale, 'New Art, New World', with his fellow undergraduates, the future publisher Charles Booth-Clibborn and the curator Greville Worthington. With staggering chutzpah, Jopling went to New York and persuaded Jean-Michel Basquiat, Julian Schnabel and Keith Haring to donate work, which was exhibited at Jack Barclay's Rolls-Royce London showroom. This experience gave him the confidence to approach artists, and he realised 'that I like the company of artists very much'.[17] This was crucial, as living and partying with them would be part of his successful formula.

The most important meeting in Jopling's life happened in 1991 at Karsten Schubert's gallery and continued in a rum bar, when he met Damien Hirst. The artist began explaining his plans to an excited Jopling, who immediately wanted to find ways of making them happen. Sharing an instant chemistry, they found that 'we both came from Yorkshire, and both lived in Brixton'.[18] Above all, they were ambitious and wanted to get things done. Jopling told the author that when they met, they talked until 4 a.m., and 'when I told him that I had a Judd and Flavin at home, Damien said he was going to come and see me the next morning'.[19] At 9 a.m. Hirst was on the doorstep with his plans for cabinets, sharks and spots. As Jopling put it, 'Damien had his entire creative repertoire in his head, and it was for me to help him realise them. I was excited by his creativity.'[20]

It is sometimes said that Damien invented Jay, but this misses

* Jopling bought his first work of art from d'Offay aged fourteen – a limited-edition Gilbert & George book that cost £16.

the point – their alliance was necessary for Hirst to get on with his work, since very early on, Jopling became a facilitator to his creative process. An example of this is the fish cabinet, which came before Hirst's shark. Jopling recalls that, 'In 1991, Damien and I went down to Billingsgate looking for stuff for his fish tank. We had been up all night and decided to go to the morning market to find lots of different kinds of fish. We soon had a car full of stinking fish and left the market as City workers were arriving, with Damien leaning out the window singing "Heigh ho, Heigh ho, it's off to work I go". This was the whirlwind that I got caught up in with Damien.'[21] The expedition was a prelude to what is one of the most important collaborations between artist and dealer.

The next project would require a backer. Damien Hirst's shark, *The Physical Impossibility of Death in the Mind of Someone Living* (1991), began as a drawing and required considerable outlay in time and money to realise. Jopling comments, 'We made the shark piece together. We put up a bounty notice in Eastern Australia to find one and then I went to Charles Saatchi to go and get the money, who agreed to fund it to the tune of £50,000.'[22] The amount barely covered the costs of the hunter, taxidermist, freezer and other specialist apparatus, and it is doubtful whether the artist received more than a fifth of the sum. What is clear, however, is that Jopling was involved in the creation of one of the most iconic works of art of our time.

When Saatchi mounted his first YBAs exhibition at Boundary Road in 1992, it was the shark that drew everyone's attention. Nobody was more impressed than Larry Gagosian, who told the author, 'When I saw the shark for the first time, I thought this is the most impactful encounter I have ever had with a work of art. I rang Charles and asked: "Who is the artist?" Saatchi said, "Damien Hirst. He's going to be the most famous artist in the world." And he was right. Hirst was the first British artist I bought.'[23] As Michael Craig-Martin wryly observed, 'the only way an English artist can be successful is to be successful somewhere else'.[24] The shark was the beginning of Hirst's international stardom.* The 1992 Boundary Road show was to define the generation which, apart from Hirst, included Sarah Lucas, Jenny Saville and Rachel Whiteread. Saatchi had bought works for the show directly from the artists, from Karsten Schubert (works by Gary Hume, Glenn Brown, Michael Landy and Rachel Whiteread) and Jay Jopling.

In this fast-moving narrative, Jopling found his first gallery space the following year. Gerard Faggionato of Christie's introduced him to the firm's CEO, Christopher Davidge; 'I asked him for a space at 44 Duke Street [a building owned by the auction house] after I gave him a good lunch. "I want to bring contemporary art into your orbit," I told Davidge.'[25] Davidge took the bait and gave Jopling five years rent free – it was an inspired move, and from it White Cube Gallery was born. 'Here,' says Jopling, 'I was to put on many first European shows, such as the American, Richard Prince. I put on seventy-five shows in eight years there.'[26] During the first year, the five shows he staged included Tracey Emin's ironically titled *My Major Retrospective 1963–1993*, which shows that humour was always part of the YBA armoury.

A chance conversation led to the most electrifying moment in the YBAs' history. Norman Rosenthal had a gap in the Royal Academy exhibition diary and Saatchi suggested lending works from his collection. In the autumn of 1997, they mounted *Sensation*, with 100 works by forty-two artists, eighteen of them former Goldsmiths students. The show's intended shock effect caused the resignation of four venerable academicians and gave rise to acres of print, both for and against the exhibits. Described

* In 1996, Gagosian first exhibited Damien Hirst in New York with an exhibition entitled *No Sense of Absolute Corruption*, featuring two formaldehyde-preserved halves of a pig in tanks.

by some as a freakshow, it unquestionably established the YBAs as an international force. Nearly 300,000 people visited *Sensation*, and Rosenthal claimed that London had for the first time become the world's art capital. The epitome of Cool Britannia, the show was a triumph for Saatchi and Rosenthal but also for Jopling. He had handled twenty-seven of the pieces in the exhibition, including the nine works by Damien Hirst and Marcus Harvey's *Myra* (which probably caused the most offence in its depiction of the Moors murderer), as well as three works by Jopling's future wife, Sam Taylor-Wood. In Craig-Martin's view, *Sensation* was the pivot between a BC and AD period. Although for the public it was the beginning of their engagement with the YBAs, in creative terms it was more a finale. Many of the artists' best creativity was accomplished and Saatchi's collecting would never again be as potent.

Shortly after *Sensation*, Saatchi constructed a public test of the market's faith in his artists. In April 1998, Christie's held a major sale of his collection, which included American, European and recent British art, in their short-lived warehouse in Clerkenwell. Although the dealers had huge misgivings, Saatchi's faith was justified, and the sale was an enormous success, firmly establishing artists like Jenny Saville, who had no previous auction form. Saatchi was, however, criticised for the effect his dispersals had on artists' prices, giving rise to the mantra that he was more a dealer than a collector, which was increasingly true. The sale marked a step change towards the auction houses' move from an exclusively secondary market to the primary market, breaking down the last bastion of the dealer's advantage over the auctioneers. It would take a further step before this change was complete and Damien Hirst's final disruption took place at Sotheby's. Although outside the timeframe of this book, it is worth describing to round off the story.

One morning, Oliver Barker, Sotheby's contemporary expert, was passing Notting Hill Gate, where, from the top of the 94 bus, he saw the dismantling of the interior of Hirst's short-lived restaurant, Pharmacy, an enterprise based less on food than PR. Sensing an opportunity, Barker telephoned Frank Dunphy, Hirst's business manager, whom he had never met, explaining that there was potentially something exciting to be done with the restaurant's fittings and artworks. Auction houses did not hold single sales of living artists and, at the time, Hirst had a deep distrust of them, comparing the auction process to 'going to a second-hand shop to buy your clothes'. On the other hand, no dealer had the setup to sell all the larger items from Pharmacy in one go. Barker pitched the idea that Pharmacy and Hirst could be part of the great tradition of celebrity auctions, of which a famous recent example had been the 1998 Duke and Duchess of Windsor sale at Sotheby's New York. When the Pharmacy sale was announced, Hirst's dealers were anxious that it might saturate the market. As it happened, the sale doubled its estimate, making a total of £10.3 million. This success encouraged Hirst to do something far more ambitious, and 'Beautiful Inside My Head Forever' was held at Sotheby's four years later. Taking place in 2008, against the dramatic backdrop of the Lehman Brothers crash, the auction of 218 items, nearly all straight from the artist, bypassed the dealers and sold directly to the public. Asked about the event, Hirst's teasing comment was, 'It's a very democratic way to sell art and it feels like a natural evolution for contemporary art.' With 'Beautiful Inside My Head Forever', his disruption of the art market was complete.

Chapter 30

Commission Fixing at Sotheby's and Christie's

'People of the same trade seldom meet together, even for merriment and diversion, but the conversation ends in a conspiracy against the public, or in some contrivance to raise prices.'[1]

Adam Smith

By the end of the 1990s, the art market, like the stock market, was flourishing both in New York and London. The astonishing success of the 1997 sale of the Duke and Duchess of Windsor's Paris house contents at Sotheby's New York demonstrated well the booming market. The firms' two ambitious CEOs, Dede Brooks at Sotheby's New York and Christopher Davidge at Christie's London, were consolidating their power base and demonstrating a level of financial ambition for their firms that was typified by Wall Street during the period. Like others in that pursuit, they crossed the line into malfeasance. As a result, the two auction houses were rocked by the greatest scandal in their history, a commission-fixing drama which would see Alfred Taubman, Sotheby's controlling shareholder, go to jail, and both houses heavily penalised by class actions.

In 1993, the much-respected Lord Carrington was anxious to step down as part-time chairman of Christie's International.

In his place, Sir Anthony Tennant, the City grandee and chairman of Guinness who had international business experience and collected Modern British paintings, was appointed. 'I took an instant dislike to him,' Christopher Davidge recalled, finding him 'arrogant, self-opiniated and totally charmless'.[2] For his part, Tennant, who had several other city directorships to attend to, thought Davidge parochial, excluded him from meetings and neglected to keep him informed. Rather to Davidge's surprise, however, Tennant did ask about his contacts with Sotheby's, a notion which up to that point had always been taboo. Lord Hindlip, Christie's UK chairman, was not enamoured by Tennant either, and the dysfunctional relationship between the three would become a factor in the brewing scandal.

While Christie's London management cooperation was falling apart, Sotheby's New York was suffering the final mutation of its CEO, Dede Brooks, into a despot. In 1996, her autocratic style caused the loss of two key department heads, David Nash in impressionists and his wife Lucy Mitchell-Innes in contemporary, and Christie's drew level with Sotheby's on turnover for the first time since the 1950s. Craving the lifestyle of her clients, with their limousines and private jets, Brooks was always searching for ways of increasing the bottom line. The shortage of top-quality goods and the ferocious competition to secure them with reduced commissions rates led, as Philip Hook remembers, to 'much head shaking: "We can't go on like this", as people in the auction houses muttered. "Something's got to be done". And done it was. In secret, with ultimately disastrous results.'[3]

In 1993, the year he was appointed chairman, Anthony Tennant bumped into Alfred Taubman at the Royal Academy in London and asked, 'You wouldn't mind if I called you, would you, and maybe we could get together?'[4] Their first meeting was held at Taubman's flat in South Audley Street at 8.30 a.m. on Wednesday 3 February 1993. Over breakfast, Taubman attacked

Christie's business tactics. The next breakfast followed in April, setting the pattern and, one might say, the countdown to disaster. Davidge was initially anxious about Tennant's contacts with Sotheby's, but the latter made it clear that it was important to have top-level dialogue to discuss matters of mutual interest. He had made a career out of profitable relationships called 'competitive alliances' in the drinks industry, particularly in the Far East. These typically resulted in agreements like 'I will leave you alone in one country if you leave me alone in another'. For him, 'a close relationship with one's competitor is to everyone's advantage. It takes out a level of competition, which is unnecessary.'[5] In America, however, any such form of collusion was illegal.

Acting on this view, Tennant and Taubman agreed that Davidge and Brooks should be told to get together. Nobody knows exactly what they expected the chief executives to do, but they took up the brief to meet and do something. Davidge was wary and covered his back by speaking confidentially to three colleagues on an individual basis: Christopher Burge as head of Christie's New York, Lord Hindlip and François Curiel, head of jewellery. 'I believed the best form of safety was to tell my senior colleagues, so they were just as responsible as I was,' Davidge was to say later.[6] Curiel recalled, 'He told me *some* things but not enough to know the extent of what he was doing. He very cleverly parcelled out information – enough to make me feel very, very important.'[7]

The first meeting of the two CEOs was in the autumn of 1993 at The Stafford, a hotel just off St James's. Brooks and Davidge both felt awkward. They discussed guarantees, commission rates, clients with bad debts and much besides. The central issue was to improve profitability: 'I think we're all concerned about the bottom line,' Brooks said; 'we're killing each other, so we should find a way so we get paid fairly for what we do so we can provide a decent return to our shareholders.'[8] Shareholders were a big theme for Brooks, since her salary and bonus were linked to their dividends and company profits, and she appeared to put 'shareholder comfort' above client satisfaction or staff contentment. Brooks and Davidge met several times, including at JFK airport on 8 February 1995, in the back of a limousine in the short-term parking lot, where they exchanged papers and propositions. Out of this came a new commission structure aimed to increase their annual profits by up to $18 million.[9] Never was Adam Smith's precept about people of the same trade conspiring together more accurate. The Sotheby's board immediately awarded Brooks a further 20,000 stock options. She began to think that she alone could save the firm, making decisions without consulting colleagues.

On 9 March 1994, Brooks was in London chairing a meeting of the international executive committee early in the morning, when suddenly Diana Phillips, head of press, burst into the room with a press release from Christie's, announcing the new tariffs. The arrival of the press release during the meeting seemed an odd coincidence, but nobody suspected the degree of collusion involved. Perhaps the most unusual thing Brooks did that day was to invite the opinion of all those in the room, which was quite unlike her. Sotheby's soon followed suit with a price increase.

The strategy worked in economic terms, and by 1995 the two firms were doing extremely well, with a 14 per cent increase in auction sales and a 32 per cent increase in pre-tax profit. Davidge and Brooks were pleased with their achievements and, for the time being thought themselves invulnerable. The first sign of trouble came in London and may have been instigated by the art trade. In June and July 1996, two letters from the UK Office of Fair Trading voiced concerns at commission rates previously negotiable by dealers. The CEOs were rattled, and Brooks called Davidge to make their story consistent. To her surprise, he told her (erroneously) that he had kept no record of their meetings or the dates, and they agreed not to meet again. Despite this, in June 1997 Carol Vogel of the *New York Times* broke the story that the US Justice Department had subpoenaed Sotheby's and Christie's and the Antitrust Division was looking at the possibility of anticompetitive practices in the fine art auction industry.[10] The London art trade showed no surprise at the news. 'It never for a moment occurred to me,' noted Alex Wengraf, 'that Christie's and Sotheby's hadn't come to some arrangement.'[11]

By 1997, Davidge was in all-out protection mode and persuaded Lord Hindlip to give him a letter dated December 1997, in which he wrote 'to reassure you that in the unlikely event that it should happen that you are forced to resign your position because of the Antitrust hearings in the US, Christie's will fully protect your position as per your contract'.[12] Davidge had a three-year contract of $480,000 per annum, and if he was to resign, there would be a golden handshake of $1.4 million. He felt reassured, 'I had Charlie's letter, so the worst that I thought could happen to me was having to resign from Christie's.'[13]

On 5 May 1998, Christie's announced that François Pinault, the French businessman and art collector, had paid an undisclosed amount to acquire Joe Lewis's 29 per cent stake in the company, which made him the controlling shareholder. After 232 years of existence, Christie's was now a French firm. The first thing Pinault learned about his new acquisition was the long-standing antipathy between Davidge and Hindlip. The latter gave an ultimatum, 'It's either Davidge or me. I'm not staying if he does.' He spoke from a position of strength, having recently organised a series of very successful auctions, most spectacularly for the Austrian branch of the Rothschild family which had made $89.9 million and was described in the *New York Times* as 'one of the most successful sales in the history of auctions in Europe'.[14]

As a result of Hindlip's demand, Pinault asked Davidge to leave. Philip Hook takes up the story: 'Davidge responded with his bombshell: "Sack me, and the illegal collusion will come out." What illegal collusion? "The commission rates I fixed with Sotheby's." You did what? Horrified, Christie's raced to the US Department of Justice to get their confession in first and ensure their own immunity from prosecution.'[15] Davidge was given a £5 million pay-off, which Hindlip thought was outrageous, although it was largely his own doing. Sotheby's was fined $45 million in the USA. Brooks was given three years' probation and six months' home detention, as well as a criminal fine of $350,000. Alfred Taubman pleaded not guilty but was sent to federal jail for ten months and paid a $7.5 million fine. Anthony Tennant was unable to enter America again for fear of being prosecuted.

Ed Dolman, the new managing director of Christie's, reflected that 'however bad it will be for us, it will be *much* worse for Sotheby's'.[16] The first inkling that Christie's might not get off scot-free, however, came when a director from Christie's New York called a client, Herbert Black, who asked 'how much are you going to pay me? … I feel there is money owing.'[17] Black initiated a lawsuit in which he rapidly gained support. By 20 April 2000, the American courts agreed to give class action status to the suits filed by clients of Sotheby's and Christie's. Both firms ended up having to pay back $256 million, the equivalent of five years of annual pre-tax profits. For Pinault, this was a shock. Costs were slashed, staff sacked and the companies raised their buyer's premium. For the staff it was a traumatic time, and Sotheby's share price plunged. Despite this, the recovery from the scandal was remarkably fast and the extraordinary thing was how little long-term damage the case did to either firm, whose strength lay in the specialist departments that nobody blamed for the debacle.

Chapter 31

At the Millennium

Since the 1958 Goldschmidt sale, the international art market was dominated by the duopoly of Sotheby's and Christie's operating out of London and later New York. Thanks to Goldschmidt, Peter Wilson had established London as world leader, and it held that position until the mid-1980s when it was overtaken by New York (especially in the field of modern art). Some auction markets decamped elsewhere, the Chinese to Hong Kong, tribal art to Paris and Brussels, and jewels to Geneva, but London remained a formidable force with superb expertise, services and the increasingly important contemporary artists who helped keep the city interesting to collectors, curators and dealers. The rise of contemporary art gave back some of the ground dealers had lost to auctioneers over the previous half-century, and around the millennium London became host to a group of international mega-dealers. The auctioneers' ascendancy was challenged by dealers' control over the contemporary market. In the aftermath of the Sotheby's and Christie's commission-fixing scandal, a more plural market emerged in which auctioneers became dealers and dealers began to participate in the risks associated with giving guarantees at auction.

The financial penalties imposed on Sotheby's and Christie's meant that they now sought profits elsewhere. The first victim of this shift was the loss of the middle market, as thresholds for accepting property for sale rose even higher. This led to the closure of many smaller London departments like European ceramics and glass. The most profound change, and one that was already brewing during the 1990s, was the auction houses' interest in private sales. Staff were given a percentage of these sales, which could double an ambitious employee's salary. Deals that thirty years earlier would have been done by Agnew's were now being done by Sotheby's and Christie's. In their search for additional revenue, this was by far the most profitable development since the implementation of the buyer's premium. By 2000, Christie's sold $185 million worth of private sales which rose to $1.2 billion by 2013.

Dealers were appalled by this encroachment on their territory, but, paradoxically, once they got over the first shock, private sales brought a much closer cooperation between the two sides who were often partners in the same deal. A more opaque financial involvement consisted of dealers giving guarantees on major lots. This increasingly took place in New York on the most expensive lots in the impressionist, modern and contemporary sales, as vendors or their agents demanded guarantees on their Picassos and Van Goghs. Guarantees were the only way to counter a cash offer from dealers and had the potential for large overage profits but the guarantees went back a long way and until this point, auction houses had always carried the risk. However, as the process became almost *de rigueur*, no auction house had the financial muscle to accept what might add up to $500 million worth of risk in one sale. The top dealers were the natural partners with whom to spread risk and guarantee by irrevocable bid or 'IB', but in 2000 this still lay in the future.

Some saw the kind of changes that had taken place in the art market since the 1980s as echoing similar shifts in the luxury goods and fashion industry, which moved from small designers to the famous corporate-branded companies. It was not for nothing that fashion billionaires were eyeing up the auction houses, resulting in Pinault's purchase of Christie's in 1998. The following year, LVMH, belonging to Bernard Arnault, the French luxury goods tycoon, bought Phillips, the third London auction house, for $125 million. He had already tried unsuccessfully to acquire Sotheby's for many times that figure. As Simon de Pury, the former chairman of Sotheby's Europe, put it, 'Arnaud decided to create his own Sotheby's from the building blocks of Phillips, a fusty London auction house that had a long history but no pizzazz.'[1]

Arnault chose de Pury to be his instrument of change at Phillips, flattering him by saying 'you need a stage', and putting him back in the rostrum where he was a star. De Pury described the new Phillips as 'the anti-Sotheby's, the Animal House of the art world, thrilled to be driving the uptight people crazy'.[2] De Pury's stated intention was to turn a dowager into a go-go girl, and Phillips was to be elite, glamorous and contemporary. In reality, however, de Pury's reckless strategy was simply to offer more daring guarantees than Sotheby's and Christie's – as on the sale of the dealer-collector Heinz Berggruen. Once Phillips's losses mounted to several hundred million dollars, Arnault's shareholders rebelled. De Pury and his partner, Daniella Luxembourg, briefly acquired the auction house themselves, before it changed hands again and became what it is today, an auction house specialising in jewellery, contemporary art and photography. Meanwhile, Bonhams enjoyed a revival as the only London auction house that still offered the full range of departments and catered to the middle market.

Technology bookends the period covered by this book. At the beginning, the 1956 arrival of the jet aeroplane service to London and improvements to international telecommunications were crucial components of Peter Wilson's creation of an international auction house at Sotheby's. The demise of Concorde's supersonic travel to London in 2003 meant that collectors could no longer make a convenient day trip to bid at auction. At the same time, the internet meant that buying art expanded beyond New York and London to salerooms all over the world. When in theory a small auction house in Malmö could have as well-designed a portal as Sotheby's, regional auctioneers prospered, multiplying the number of clients viewing their sites online. As far as the old master dealer Alfred Cohen was concerned, it took some of the fun out of the game: 'It was no longer possible to make discoveries in the same way as there were now large numbers of potential buyers looking at obscure sales.'[3]

For dealers, the internet and high-resolution digital images changed the way they did business, leading to the demise of the traditional Bond Street gallery, the flight to upper floors (with the exception of contemporary art) and the rise of art agents. When Harry Blain started dealing in prints in 1992, he recalls that he still used a very old-fashioned way of reaching clients, sending out 10 x 8-inch transparencies costing about £50 each. Over the next decade, the speed, range, and costs came down, offering a much more immediate and international clientele, who were now seconds rather than a week away from receiving images. It wasn't just contemporary dealers who benefited. For old master dealer-agents like Simon Dickinson, it was no longer necessary to have a shop window and hold stock. His successful business model was based on sending high-resolution images to forty or so of the great collectors of the world by email. Acting as an agent had further advantages in the internet age, since it meant that the price was kept secret and no publicity ensued in the event of a failure to sell the item.

The internet brought other changes, putting more power in the hands of the public often at the expense of dealers. With Artnet, all auction pricing information was available at the tip of a finger. To misquote Oscar Wilde, everybody now knew the price of everything and the value of nothing. This is echoed by Ivor Braka's belief that new buyers would do anything rather than give a dealer profit, an attitude that led to their growing preference for auctions. Richard Green was one of the few dealers who remained unconcerned, confident in creating a new price level, irrespective of what he paid at auction. He had become his own brand.

Sotheby's and Christie's took longer to understand how to harness the power of new technology and in doing so they went all the way from a 'gentlemen's market' to a virtual one. Used to dealing with clients on a regional basis, the companies' systems were based on that principle. Their new clients no longer conformed to this pattern. Before the millennium, auctioneers knew who nearly all the clients were, or were likely to be, but, as Thomas Seydoux from Christie's put it, now 'we don't even know them, and they come in at the top from the start'. This meant much more stringent bidding arrangements. James Roundell recalls that, 'There were no paddles until 1998; the bidders held up their hands, and the auctioneer called out the name of the individual, if it was an unknown person then a staffer would get their name.' With paddles, registering meant handing over bank details, financial guarantees and much else. After hundreds of years of practice, a hand raised at auction was no longer sufficient.

One important London initiative was the Art Loss Register founded in 1990, a collaboration by art insurers, dealers and the main auction houses. The majority owner was the founder, Julian Radcliffe, a former Lloyd's underwriter, and Sotheby's and Christie's were minority shareholders. The register rapidly became the world's most comprehensive database on lost, stolen and at-risk art and antiquities. Used for due diligence, it dealt with 400,000 enquiries a year. Nobody, however, imagined at the time that its principal public activity would be restitution of works of art looted from Jewish collections during World War II.

Around the millennium, Restitutions appeared as a new source of business. The driving force was the 1998 Art Restitution Act, which forced Swiss banks, among other institutions, to reveal, trace or return objects stolen in World War II. The process was given further impetus by the fall of the Berlin Wall and the opening of the East German archives. The Jewish lobby in New York had a strong sense of unfinished business about righting the wrongs of Nazi persecution, and a world of researchers and agents operating on a no-win no-fee basis grew up to track down objects on behalf of claimants. Although claims often had little success in a court of law, they were felt very strongly in the court of public opinion. As a result, museums, collectors, dealers and auction houses became concerned about their reputations in owning and trading looted or tarnished art. Restituted art produced a considerable body of sales for the salerooms over the next twenty years, and any item without a cast-iron provenance between 1933 and 1945 was automatically the subject of contention and consequently unsaleable without an agreement with the claimants.

By 2000, the London art trade may have passed its golden age, which ran from 1958 until about 1987. It still enjoyed a silver age with its many world-class dealers, like Giuseppe Eskenazi in Chinese art and Danny Katz in sculpture, to mention only two. London had also transformed from the old master capital of the world to a major contemporary art city with Tate Modern, great dealers in the field and some of the world's most sought-after artists. As Anthony d'Offay later claimed, 'I would say that London is the second city in the world for contemporary art, but the range of galleries in London now rivals even New York.' With the patronage of billionaire art collectors, the power and importance of the independent curator and art advisor grew.

Art had become an international circus promoted by fairs, and was seen by many as the new rock'n'roll. Exhibitions were masterminded by celebrity curators like Hans Ulrich Obrist, appointed co-director of the Serpentine Gallery in 2006, and voted the most powerful man in the art world three years later.[4] Frieze London, held in October, became the focus of this arena, with auction houses running their sales alongside the fair rather than the other way around, as would have been the case twenty years earlier. With the huge expansion of fairs, dealers and galleries, the holding of a large stock became secondary, and fairs became a platform to meet and sell to new collectors.

The new emphasis on contemporary was timely, since goods were running out in traditional areas. Important old masters were harder to find, as country houses had cut down their collections over the previous fifty years. During the 1970s, Sotheby's held thirteen old master auctions a year, over double the number today. Discoveries could still be made, however, and London could still pack a heavy punch using its expertise. When George Gordon of Sotheby's identified a previously misattributed Rubens *Massacre of the Innocents* in 2002, it was sold for £49.5 million to Sam Fogg, bidding on behalf of Ken Thomson. Rubens's violent subject matter, a likely impediment in the 1980s heyday of soft-focus Renoir, did not put anybody off in the YBA era. Fernand Leger's battle cry, 'the pretty is the enemy of the beautiful', had become true across the market.

Writing in 1990, the dealer Charles Riley noted that, 'The long and distinguished reign of London as the art world's centre may be nearing its end. Undisputed capital of the auction world until a decade ago, still the sentimental favourite among the trade's dominant experts, a considerable number of whom call it home, London is suffering from a lack of local buying power and the steep cost of doing business for galleries that have already taken on considerable debt.'[5] Indeed, until the end of the twentieth century, the art market was dominated by the

USA, which had approximately a 40 per cent share of global art sales in 1990, in comparison to the UK which had around 25 per cent.*

To Georgina Adam of *The Art Newspaper*, 'The democratisation and popularisation of art from the 1960s onwards seems now to have reversed. To an extent today there is an increasing identification of art, artists and art galleries with the global 0.01 per cent.'[6] James Holland-Hibbert identified Sotheby's, Christie's, Gagosian, Pace, Hauser & Wirth, David Zwirner and Acquavella as 'the Big Seven', and the gap between these superstars and the rest of the market became ever wider. Indeed, this divergence in the market replaced the pyramidal structure of the past, which went through all the gradations of runners, traders and provincial dealers up to Bond Street. Outside the rarefied field of contemporary art, however, this split is less evident, and the internet is alive with buyers and sellers who form the base of a pyramid which no longer has a national home. This activity demonstrates the extent to which the art world, and the way it does business, has evolved across the half-century covered in this book.

If London has continued to thrive as a centre of the art market in the first part of the twenty-first century it is because of its attraction to Arab sheiks, oligarchs and the other super-rich of the world. Much was made of its convenient position halfway between the time zones of New York and Hong Kong. During the first decade of the new millennium, London occupied a position as the leading city of Europe, although by 2020 Brexit has somewhat dented this image.† Despite this, London

* Statistics of this kind are difficult to find and often unreliable. These figures come from Adam, 2017, p. 134. They have been adjusted by the art-world statistician Jeremy Eckstein. The figures for other countries are too unreliable to quote.

† Larry Gagosian told the author that, as far as he was concerned, 'Brexit is a bummer'.

is and always has been in a state of constant reinvention. As Jan Morris observed, 'She is a city that moves in cycles. She is an island capital, an offshore meeting-place or conduit, through which ideas and influences pour in successive waves … emerging from one cycle she will presently enter the next.'[7] There is no way of telling what that will be.

Acknowledgements

The germ of the book was a chance conversation with a friend and neighbour, Colin Sheaf, ex-Christie's and Bonhams. The trade reacted with great enthusiasm and I am very grateful to them all for giving me such interesting interviews: Peter Adler, Brian Allen, Julian Agnew, Charles Avery, Jon Baddeley, George Bailey, Godfrey Barker, Oliver Barker, Julian Barran, John Baskett, Lee Beard, Alex Bell, Jörg Bertz, Roger Bevan, Ivor Braka, Ben Brown, Anthony Browne, Humphrey Butler, Juliet Byford, the late Anthony du Boulay, Vanessa Brett, Harry Blain, Nick Bonham, Richard Calvocoressi, Guillaume Cerutti, Charles Booth Clibborn, Iona Bonham-Carter, Jonathan Clark, Hugo Cobb, the late Alfred Cohen, Desmond and Alexander Corcoran, Cathy Courtney, Michael Craig-Martin, Emma Crichton-Miller, Miranda Crimp, Emma Cunningham, Caroline Cuthbert, Richard Davenport-Hines, Alastair Dickenson, Simon Dickinson, Patrick Dingwall, Martin Drury, Dendy Easton, James Ede, Annamaria and Victor Edelstein, David Ellis-Jones, Lance Entwistle, John Erle-Drax, Giuseppe Eskenazi, Maggie Evans, James Faber, Richard Falkiner, Oliver Forge, Anton Gabszewicz, Larry Gagosian, Edward Gibbs, Thomas Gibson, George Goldner, Michelle Gower,

Cornelia Grassi, Bennie Gray, Richard Green, Bernard de Grunne, Johnny van Haeften (who gave me the run of his press cuttings file and archive), Julian and Barbara Harding, Julian Hartnoll, Thomas Heneage, Katrin Henkel, Richard Herner, Lord Hindlip, Peter Hinwood, James Holland-Hibbert, Peter Holmes, Peter Johnson, William Joll, Kasmin, Danny Katz, Hilary Kay, Robin Kern, Roger Keverne, Christopher Kingzett, James Kirkman, Chris Kneale, Richard Knight, Dominic Jellinek, Jay Jopling, Marie-Louise Laband, Alastair Laing, Mark Law, Simon Lee, Martin Leggatt, Martin Levy, Christopher Lewin, Thomas Lighton, Marcus Linnell, Nicholas Logsdail, Gilbert Lloyd, John Lumley, Katherine Maclean, Richard Madley, Rupert Maas (who kindly showed me the manuscript memoir by his father and that of Christopher Wood), Louise Malcolm, Huon Mallalieu, John Mallet (who kindly made his unpublished memoirs available to me), Errol Manners, Patrick Matthiesen, James Mayor, Alex Meddowes, David Messum, James Miller, Liz Mitchell, David Mlinaric, Lisa Mondiano, Susan Morris, Richard Morphet, Anthony Mould, David Nash, Sir Philip Naylor-Leyland, Angela Nevill, Rena Neville, Geraldine Norman (who gave me the run of her extensive archive), Nicholas Norton, Judith Nugée, Anthony d'Offay, Barbara Paca, Frank Partridge, Ray Perman, Anders Petterson, Simon Phillips, Susie Pollen, David Posnett, Julian Radcliffe, James Rawlin, Howard Ricketts, Sir Hugh Roberts, Orlando Rock, Yvonne Robinson, Pamela Roditi, Francis Russell, Diana Scarisbrick, Tim Schroder, Nick Serota, Lewis Smith, Peyton Skipwith, Stephen Somerville, Anthony Speelman, Simon Spero, Simon Stock, Martin Summers, Lanto Synge, Michael Thomson-Glover, Flora Turnbull, Robert Upstone, Lara Wardle, Hermoine Waterfield, Heather Waddell, Peter Waldron, Jeremy Warren, Pat Wengraf, Martin Westgarth, John Whitehead, Elizabeth Wilson, Philip Wright, Henry Wyndham, Bill Zachs and Rainer Zietz.

Jonathan Harris set up the project by opening his address book to me in the early stages and providing excellent advice on inclusion. I owe a special debt to John Culme, who provided so much glorious material for the chapter on silver. Those who read the entire manuscript and improved the book beyond measure were Noël Annesley, Philip Hook, David Moore-Gwyn and Martin Royalton-Kisch. Further thanks to my agent, Georgina Capel, and my editors, Richard Milbank, Gina Blackwell and Ellie Jardine. My greatest debt is to Adrian Eeles, who acted as a director of studies, and Dr Verity Mackenzie, for her excellent suggestions and support.

Select Bibliography

Adam, Georgina, *Dark Side of the Boom: The Excesses of the Art Market in the Twenty-first Century* (London, 2017).

Agnew, Geoffrey, *Agnew's 1817–1967* (London, 1967).

Agnew, Julian, and others, *A Dealer's Record, Agnew's 1967–81* (London, 1981).

Agnew, Julian, and others, *Agnew's 1982–1992* (London, 1992).

Barbican, *Postwar Modern: New Art in Britain, 1945–1965* (London, 2022).

Battie, David (ed.), *The Antiques Roadshow* (London, 2005).

Berthoud, Roger, *The Life of Henry Moore* (London, 1987).

Birch, James, *Bacon in Moscow* (London, 2022).

Bohm-Duchen, Monica (ed.), *Insiders Outsiders: Refugees from Nazi Europe and their Contribution to British Visual Culture* (London, 2019).

Boulay, Anthony du, *The Duberly Collection of Chinese Art at Winchester College* (Winchester, 2019).

Bowron, Edgar Peters (ed.), *Buying Baroque: Italian Seventeenth-Century Paintings Come to America* (Philadelphia, 2017).

British Art Market Federation, *The British Art Market 1997: A study of the value of the art and antique market in Britain and the implications of EU harmonisation of import VAT and Artists Resale Rights* (London, 1997).

Browse, Lillian, *Duchess of Cork Street: The Autobiography of an Art Dealer* (London, 1999).

Cecil, Mirabel, and David Mlinaric, *Mlinaric on Decorating* (London, 1988).

Christie's, *Going Once: 250 Years of Culture, Taste and Collecting at Christie's* (London, 2016).

Cooper, Douglas (ed.), *Great Private Collections* (London, 1963).

Cooper, Douglas, *Alex Reid & Lefevre 1926–76* (London, 1976).

Cooper, Jeremy, *Under the Hammer: The Auctions and Auctioneers of London* (London, 1977).

Cooper, Jeremy, *Dealing with Dealers: the Ins and Outs of the London Antiques Trade* (London, 1985).

Coppet, Laura de, and Alan Jones, *The Art Dealers* (New York, 2002).

Davids, Roy, and Dominic Jellinek, *Provenance: Collectors, Dealers and Scholars: Chinese Ceramics in Britain & America* (Oxfordshire, 2011).

Day, Richard, *Artful Tales* (London, 2008).
Eeles, Adrian, 'Memoir of the Print Trade', *Print Quarterly*, vol. XXXVIII, no. 4, December 2021.
Eskenazi, Giuseppe, *A Dealer's Hand: The Chinese Art World through the Eyes of Giuseppe Eskenazi* (London, 2012).
Faith, Nicholas, *SOLD: The Revolution in the Art Market* (London, 1985).
Feaver, William, *The Lives of Lucian Freud: Youth 1922–68* (London, 2019).
Felch, Jason, and Ralph Frammolini, *Chasing Aphrodite, The Hunt for Looted Antiquities at the World's Richest Museum* (Boston, 2011).
Fox, Dan, 'Then & Now: British art in the 1990s', *Frieze Magazine*, Issue 159, 23 November 2013.
Fredericksen, Burton B., *The Burdens of Wealth: Paul Getty and his Museum* (Bloomington, 2015).
Fullerton, Elizabeth, *ARTRAGE! The Story of the BritArt Revolution* (London, 2016).
Gallagher, Ann (ed.), *Damien Hirst* (London, 2012).
Gerlis, Melanie, *The Art Fair Story: A Rollercoaster Ride* (London, 2021).
Girouard, Blanche, *Portobello Voices* (Stroud, 2013).
Goodison, Nicholas, and Robin Kern (eds.), *Eighty Years of Antique Dealing* (London, 2004).
Grant, Thomas, *Jeremy Hutchinson's Case Histories* (London, 1998).
Haden-Guest, Anthony, *True Colors: The Real Life of the Art World* (New York, 1996).
Hall, Michael, *The Harold Samuel Collection* (London, 2012).
Halliday, Stephen, *London's Markets: From Smithfield to Portobello Road* (Stroud, 2014).
Harrison, Martin, *Transition: The London Art Scene in the Fifties* (London, 2002).
Hebborn, Eric, *Confessions of a Master Forger* (London, 1997).
Herbert, John, *Inside Christie's* (London, 1990).
Herrmann, Frank, *Sotheby's: Portrait of an Auction House* (London, 1980).
Hillier, Bevis, *The New Antiques* (London, 1978).
Hindlip, Charles, *An Auctioneer's Lot: Triumphs and Disasters at Christie's* (London, 2016).
Hook, Philip, *The Ultimate Trophy: How the Impressionist Painting Conquered the World* (London, 2009).
Hook, Philip, *Breakfast at Sotheby's, An A–Z of the Art World* (London, 2013).
Hook, Philip, *Rogue's Gallery: A History of Art and its Dealers* (London, 2017).
Houghton, Arthur, *Standing Still is Not An Option: A Memoir* (no place, 2022).
Howard, Jeremy (ed.), *Colnaghi: The History* (London, 2010).
Johnson, Peter, *Heart in Art: A Life in Paintings* (London, 2010).
Keating, Tom, Geraldine Norman and Frank Norman, *The Fake's Progress: Tom Keating's Story* (London, 1977).

Keen, Geraldine, *The Sale of Works of Art: A Study based on the Times-Sotheby Index* (London, 1971).

Kingzett, Dick, *Sir Geoffrey Agnew 1908–1986 Dealer and Connoisseur* (London, 1988).

Lacey, Robert, *Sotheby's – Bidding for Class* (London, 1998).

Lennox-Boyd, Edward, *Masterpieces of English Furniture: The Gerstenfeld Collection* (London, 1991).

Maas, Jeremy, 'Memoir' (unpublished manuscript, no date).

Mallet, John, 'Memoir of his time at Sotheby's' (unpublished manuscript, no date).

Mason, Christopher, *The Art of the Steal: Inside the Sotheby's-Christie's Auction House Scandal* (New York, 2004).

McAlpine, Alistair, *Journal of a Collector* (London, 1994).

McLean, Katherine, and Philip Hook, *Sotheby's Maestro: Peter Wilson and the post-war Art World* (London, 2017).

Mellon, Paul, *Reflections in a Silver Spoon, a Memoir* (New York, 1992).

Mellor, David, *The Sixties Art Scene in London* (London, 1993).

Melly, George, *Rum, Bum and Concertina* (London, 1977).

Melly, George, *Don't Tell Sybil: An Intimate Memoir of E. L. T. Mesens* (London, 1997).

Miles, Roy, *Priceless: The Memoirs and Mysteries of Britain's No. 1 Art Dealer* (London, 2003).

Mortimer, Charlie, *Lucky Lupin* (London, 2016).

Mould, Philip, *Sleuth: The Amazing quest for Lost Art Treasures* (London, 2009).

Norman, Geraldine (ed.), *Bob Hecht By Bob Hecht* (Piraeus, 2014).

O'Connell, Brian, *John Hunt, the Man, the Medievalist, the Connoisseur* (Dublin, 2013).

Parker, John, *Great Art Sales of the Century* (London, 1975).

Pury, Simon de, and William Stadiem, *The Auctioneer: a memoir of great art, legendary collectors and record-breaking auctions* (London, 2016).

Reitlinger, Gerald, *The Economics of Taste: the Rise and Fall of Picture Prices 1760–1960*, vol.I (London, 1961).

Reitlinger, Gerald, *The Economics of Taste: The Art Market in the 1960s*, vol. III (London, 1970).

Renfrew, Colin, *Loot, Legitimacy and Ownership: The Ethical Crisis in Archaeology* (London, 2000).

Robertson, Bryan, and John Russell and Lord Snowdon, *Private View: The lively world of British Art* (London, 1965).

Robinson, Duncan, 'The London Art World, 1950–1965', *Voices of Art* (bl.uk/voices-of-art).

Roland, Henry, *Behind the Façade: Recollections of an art dealer* (London, 1991).

Rosenthal, Norman, and others, *Sensation: Young British Artists from the Saatchi Collection* (London, 1997).
Russell Taylor, John, and Brian Brooke, *The Art Dealers* (New York, 1969).
Seldes, Lee, *The legacy of Mark Rothko* (New York, 1996).
Sewell, Brian, *Outsider: Almost Always Never Quite* (London, 2011).
Sewell, Brian, *Outsider II: Almost Always Never Quite* (London, 2012).
Shakespeare, Nicholas, *Bruce Chatwin* (London, 1999).
Shnayerson, Michael, *Boom: Mad Money, Mega Dealers, and the Rise of Contemporary Art* (New York, 2020).
Shortland, Anja, *Lost Art: The Art Loss Register Casebook*, vol. II (London, 2021).
Solomon, Andrew, 'Cache Barriers', *Harpers & Queen*, 1 September 1989 (http://andrewsolomon.com/articles/cache-barriers/).
Speelman, Anthony, *A Tale of Two Monkeys: Adventures in the Art World* (London, 2020).
Spira, Peter, *Ladders and Snakes* (London, 1997).
Stallabrass, Julian, *High Art Lite: The Rise and Fall of Young British Art* (London, 2006).
Stevens, Mark, and Annalyn Swan, *Francis Bacon: Revelations* (New York, 2020).
Strauss, Michel, *Pictures, Passion and Eye: A Life at Sotheby's* (London, 2011).
Stuart-Penrose, Barrie, *The Art Scene* (London, 1969).
Stourton, James, *Great Collectors of Our Time* (London, 2007).
Sykes, Christopher Simon, *Hockney: The Biography Vol I 1937–75* (London, 2011).
Synge, Lanto, *Mallett Millenium* (London, 1999).
Synge, Lanto, *Telling Threads* (London, 2015).
Tickner, Lisa, *London's New Scene: Art and Culture in the 1960s* (London, 2020).
Tinniswood, Adrian, *Noble Ambitions: The Fall and Rise of the Post-War Country House* (London, 2021).
Tonkovich, Jennifer, 'Hans Calmann and the American Market for Old Master Drawings, 1937–73', *Master Drawings* No. 59 2017, pp. 49–72.
Van der Lande, Joanna, 'The Antiquities Trade: A reflection on the past 25 years, Part 1', *Cultural Property News* (https://culturalpropertynews.org/the-antiquities-trade-a-reflection-on-the-past-25-years-part-1/).
Vyner, Harriet, *Groovy Bob: The Life and Times of Robert Fraser* (London, 1999).
Waddell, Heather, *The London Art World 1979–99* (London, 2000).
Waddell, Heather, *London Art Snakes & Ladders* (unpublished, 2010).
Waterfield, Hermione, and John C. H. King, *Provenance: Twelve Collectors of Ethnographic Art in England 1760–1990* (London, 2009).
Watson, Peter, *The Caravaggio Conspiracy: A True Story of Deception, Theft and Smuggling in the Art World* (London, 1983).
Watson, Peter, *Sotheby's: The Inside Story* (New York, 1997).
Watson, Peter, and Cecilia Todeschini, *The Medici Conspiracy: The Illicit Journey of Looted Antiquities from Italy's Tomb Raiders to the World's Greatest Museums* (New York, 2006).

Wengraf, Alex, *Memories of a London Fine Art Dealer* (London, 2020).
Westgarth, Mark, *Sold! The Great British Antiques Story* (Leeds, 2019).
Wood, Christopher, *The Great Art Boom 1970–1997* (Weybridge, 1997).
Wood, Christopher, *My Forty Years in the Art World* (unpublished, no date).
Wraight, Robert, *The Art Game* (London, 1965).

Endnotes

Preface

1 *TLS*, 13 January 1966. Review of John Russell and Bryan Robertson's *Private View* (1965), a portrait of the art world in the 1960s.

Introduction

1 Henry Roland, 1991, pp. 83–4.
2 *The Times*, 21 June 1984, p. 14.
3 Robert Wraight, 1965, p. 121.
4 Barrie Stuart-Penrose, 1969, pp. 10 and 18.
5 *Bastian*, vol. 2, Spring 2019, p. 10.
6 David Sylvester, *London Review of Books*, 30 March 2000.
7 Robert Lacey, 1998, p. 155.
8 Philip Hook, interview with the author.
9 Richard Day, 2008, p. 196.
10 *Sunday Times*, 15 May 1960, pp. 8–9.
11 Hook, 2017, p. 208.
12 Interview with the author.
13 Lanto Synge, 2015, p. 64.
14 Synge, 2015, p. 259.
15 Synge, 2015, p. 260.
16 Interview with the author.
17 See Adrian Eeles, 2021, p. 387.
18 Jeremy Cooper, 1985, p. 21.
19 Giuseppe Eskenazi, 2012, p. 58.

Chapter 1

1 Lillian Browse, 2009, p. 153.
2 Cooper, 1977, p. 61.
3 John Herbert, 1990, p. 26.
4 Herbert, 1990, p. 24.
5 Brian Sewell, 2012, p. 4.
6 Anthony du Boulay to the author.
7 Quoted Lacey, 1998, p. 95.
8 Geraldine Norman in Katherine MacLean and Philip Hook, 2017, p. 142.
9 Diana Scarisbrick in MacLean and Hook, 2017, p. 38.
10 Lacey, 1998, p. 111.
11 Lacey, 1998, p. 111.
12 Lacey, 1998, p. 114.
13 Lacey, 1998, p. 119.
14 Frank Herrmann, 1980, p. 353.
15 Herrmann, 1980, p. 341.
16 Herbert, 1990, p. 21.
17 Herrmann, 1980, p. 371.
18 Herrmann, 1980, pp. 371–2.
19 John Partridge in MacLean and Hook, 2017, p. 82.
20 Lacey, 1998, p. 124.
21 John Parker, 1975, p. 112.
22 Herbert, 1990, p. 24.

Chapter 2

1 Cooper, 1977, pp. 82–3.
2 This account comes from Michel Strauss, 2011, p. 214.

3 Christopher Wood, unpublished, n.p.
4 Julian Agnew in MacLean and Hook, 2017, p. 171.
5 Lacey, 1998, p. 152.
6 Sewell, 2011, p. 200.
7 Herbert, 1990, p. 230.
8 Herbert, 1990, p. 40.
9 Julien Stock in MacLean and Hook, 2017, p. 121.
10 Parker, 1975, p. 112.
11 Marcus Linell in MacLean and Hook, 2017, p. 42.
12 Herbert, 1990, p. 231.
13 Strauss, 2011, p. 77.
14 Strauss, 2011, p. 77.
15 Browse, 1999, p. 153.
16 Lacey, 1998, p. 150.
17 Interview with the author.
18 Lacey, 1998, p. 150.
19 Cyril Humphris in MacLean and Hook, 2017, p. 89.
20 Interview with Ray Perman.
21 Synge, 1999.
22 Christie's *Review of Season*, 1973–4, p. 17.
23 Cooper, 1977, p. 25.
24 Herrmann, 1980, p. 390.
25 Lacey, 1998, p. 166.

Chapter 3

1 Charles Hindlip, 2016, p. 239.
2 MacLean and Hook, 2017, p. 86.
3 See *The Times*, 1 December 1964.
4 Parker, 1975, p. 117.
5 Wood, unpublished, n.p.
6 Wood, unpublished, n.p.
7 Ray Perman to the author.
8 Wood, unpublished, n.p.
9 Wood, unpublished, n.p.
10 Sewell, 2012, p. 4.
11 Wood, unpublished, n.p.
12 Herbert, 1990, pp. 60–1.
13 Sewell, 2011, p. 222.
14 Wood, unpublished, n.p.
15 Wood, unpublished, n.p.
16 Herbert, 1990, p. 102.
17 Herbert, 1990, pp. 108–9.
18 Herbert, 1990, p. 110.
19 Herbert, 1990, p. 111.
20 Eskenazi, 2012, pp. 68–9.
21 Herbert, 1990, p. 160.
22 Parker, 1975, p. 87.
23 Hindlip, 2016, p. 239.
24 Peter Watson, 1997, p. 351.
25 Herbert, 1990, p. 211.
26 Lacey, 1998, p. 176.
27 Lacey, 1998, p. 174.
28 Christopher Mason, 2004, p. 70.
29 Wood, unpublished, n.p.
30 Wood, unpublished, n.p.
31 Mason, 2004, p. 71.
32 John Baskett, email to author.
33 Interview with the author.

Chapter 4

1 Stuart-Penrose, 1969, p. 34.
2 Interview with the author.
3 John Russell Taylor and Brian Brooke, 1969.
4 Interview with Christopher Kingzett.
5 Quoted Wraight, 1965, p. 122.
6 Russell Taylor and Brooke, 1969, p. 85.
7 Agnew, 1967, p. 60.
8 Stuart-Penrose, 1969, p. 40.
9 Stuart-Penrose, 1969, p. 40.
10 Interview with Christopher Kingzett.
11 Agnew, 1967, p. 58.
12 Interview with John Baskett.
13 Jeremy Howard, 2010, p. 51.
14 Sewell, 2012, p. 44.
15 Howard, 2010, p. 5.
16 Paul Mellon, 1992, pp. 279–80.
17 Lord Rothschild in MacLean and Hook, 2017, p. 220.
18 Anon interview with the author.
19 Wood, unpublished, n.p.

20 *Bastian*, vol. 2, Spring 2019, p. 12.
21 *Bastian*, vol. 2, Spring 2019, p. 12.
22 Interview with the author.
23 Interview with David Ellis-Jones.
24 Eskenazi, 2012, p. 32.

Chapter 5

1 Quoted in J. Patrice Marandel in Edgar Peters Bowron, 2017, p. 91.
2 Browse, 1999, p. 53.
3 Browse, 1999, p. 59.
4 Wood, unpublished, n.p.
5 Day, 2008, pp. 10–11. Day's version is inaccurate and I am grateful to John Baskett for giving me the correct version.
6 Interview with Alfred Cohen.
7 Julian Agnew in MacLean and Hook, 2017, p. 174.
8 J. Patrice Marandel in Bowron, 2017, p. 80.
9 Information from Julian Agnew.
10 Information from Thomas Heneage.
11 Interview with the author.
12 Herbert, 1990, p. 95.
13 Herbert, 1990, p. 95.
14 House of Lords Debate, 15 July 1969.
15 Parker, 1975, p. 90.
16 Anon interview with the author.
17 Wood, unpublished, n.p.
18 Anthony Speelman, 2020, p. 40.
19 Anthony Speelman, 2020, p. 170.
20 Anthony Speelman, 2020, p. 37.
21 J. Patrice Marandel in Bowron, 2017, p. 78.
22 J. Patrice Marandel in Bowron, 2017, p. 81.
23 J. Patrice Marandel in Bowron, 2017, p. 87.
24 Anon interview with the author.
25 Anon interview with the author.
26 Interview with the author.
27 Interview with the author.
28 Interview with the author.
29 Information from the late Joe Och.
30 Interview with the author.

Chapter 6

1 Browse, 1999, p. 119.
2 Browse, 1999, p. 110.
3 Geraldine Keen, 1971, p. 31.
4 *The Sunday Times*, 15 May 1960, pp. 8–9.
5 George Melly, 1997, p. 109.
6 Kasmin interview with the author.
7 Barbican, 2022, p. 18.
8 Quoted William Feaver vol. I, 2019, p. 329.
9 *The Sunday Times*, 15 May 1960, pp. 8–9.
10 Melly, 1997, p. 109.
11 Melly, 1997, p. 109.
12 Information from Yvonne Robinson.
13 Roland, 1991, p. 63.
14 Roland, 1991, p. 63.
15 Feaver, vol. I, 2019, p. 196.
16 *The Guardian*, 2007.
17 James Mayor 'Stories', *The Mayor Gallery*, https://www.mayorgallery.com/stories/.
18 Melly, 1997, p. 113.
19 Melly, 1997, p. 71.

Chapter 7

1 Russell Taylor and Brooke, 1969, p. 254.
2 Stuart-Penrose, 1969, p. 36.
3 *The Observer*, 16 October 1960.
4 Feaver, vol. I, 2019, p. 496.
5 Interview with the author.
6 *Bastian*, vol. 2, Spring 2019, p. 12.
7 Lee Seldes, 1996, p. 55.
8 *The Observer*, 16 October 1960.
9 *The Sunday Times*, 15 May 1960, pp. 8–9.

10 Russell Taylor and Brooke, 1969, p. 254.
11 Interview with the author.
12 Seldes, 1996, p. 56.
13 Interview with the author.
14 *The Observer*, 16 October 1960.
15 Interview with the author.
16 *Bastian*, vol. 2, Spring 2019, p. 12.
17 Feaver, vol. I, 2019, p. 496.
18 Feaver, vol. I, 2019, p. 504.
19 Seldes, 1996, p. 58.
20 Anthony Haden-Guest, 1996, p. 9.
21 Interview with the author.

Chapter 8

1 Quoted in Harriet Vyner, 1999.
2 Vyner, 1999, p. 99.
3 Russell Taylor and Brooke, 1969, p. 255.
4 The Tooth and Waddington galleries briefly joined up.
5 Quoted in Sylvester review *LRB*, 2000.
6 Lisa Tickner, 1996, p. 72.
7 Quoted Tickner, 2020, p. 72.
8 Christopher Sykes, 2011, p. 131.
9 Sykes, 2011, p. 139.
10 Quoted Tickner, 2020, p. 85.
11 Tickner, 2020, p. 87.
12 Vyner, 1999, p. 119.
13 Quoted Tickner, 2020, p. 79.
14 Quoted Sylvester review *LRB*.
15 Quoted Tickner, 2020, p. 77.
16 Vyner, 1999, p. 73.
17 Vyner, 1999, p. 115.
18 Vyner, 1999, p. 127.
19 Interview with the author.
20 Barbican, 2022, p. 19.
21 Russell Taylor and Brooke, 1969, p. 173.
22 Quoted in *The Times*, interview with artist, 19 September 2022.
23 Interview with the author.
24 I am grateful to Lee Beard for these notes on Ben Nicholson.

Chapter 9

1 Interview with the author.
2 See Christopher Gilbert in Edward Lennox-Boyd, 1991, p. 36.
3 Information from Martin Drury.
4 Interview with David Salmon.
5 Martin Drury to the author.
6 Synge, 2015, p. 341.
7 Synge, 2015, p. 90.
8 Synge, 2015, p. 78.
9 Synge, 2015, p. 21.
10 It was built for Colnaghi in 1912.
11 John Partridge in MacLean and Hook, 2017, p. 83.
12 Interview with David Salmon.
13 See James Stourton, 2007, p. 94.
14 Nicholas Goodison and Robin Kern, 2004, p. 190.
15 John Partridge in MacLean and Hook, 2017, p. 84.
16 Interview with the author.
17 *Daily Telegraph*, 12 June 1995.
18 Despite that, Godfrey Barker could still report a pre-tax profit for Mallett in 93/94 of £1.04 million. *Daily Telegraph*, 12 June 1995.
19 Interview with the author.
20 Charlie Mortimer, 2016, pp. 188–9.

Chapter 10

1 Interview with the author.
2 Hamish Bowles, *Vogue*, 31 July 2018.
3 Sewell, 2012, p. 183.
4 Mirabel Cecil and David Mlinaric, 1998, pp. 75–6.
5 See Christopher Gibbs catalogue, Christie's, 2006.

6 Interview with the author.
7 Anon interview with the author.
8 Mortimer, 2016, p. 199.
9 Mortimer, 2016, p. 201.
10 Interview with the author.
11 Mortimer, 2016, p. 195.
12 Mortimer, 2016, p. 175.
13 *Sunday Times*, 6 April 2008.
14 Synge, 2015, p. 85.
15 Anon interview with the author.
16 See *The Art Newspaper*, 21 May 2018.

Chapter 11

1 Interview with the author.
2 Interview with the author.
3 Cyril Humphris in MacLean and Hook, 2017, pp. 88–9.
4 Hermione Waterfield and John C. H. King, 2009, p. 155.
5 Nicholas Shakespeare, 1999, p. 180.
6 Shakespeare, 1999, p. 182.
7 Anon interview with the author.
8 Shakespeare, 1999, p. 185.
9 Interview with the author.
10 Interview with the author.
11 Waterfield, p. 135.
12 Adenike Cosgrove, 'Tits and Pricks: About Hermione Waterfield and the Creation of the 'Tribal Art' Department at Christie's', *ÌMỌ̀ DÁRA*, 2018.

Chapter 12

1 Email to the author.
2 Email to the author.
3 Information from Errol Manners.
4 Herbert, 1990, p. 44.
5 *The Argus*, 10 July 1969, p. 12.
6 The Metropolitan Museum New York curator.
7 Information from Errol Manners.
8 Sir John Plumb, 'The Intrigues of Sèvres', *House & Garden*, January 1986.
9 Plumb, 1986.
10 Plumb, 1986.
11 Interview with the author.
12 Interview with Mark Law.
13 *ATG*, 18 September 2021.
14 *ATG*, 18 September 2021.
15 Quoted in sale catalogue.
16 Foreword, *Simon Spero 50th Anniversary Exhibition 1964–2014: Early English Porcelain 1748–1783*.
17 Introduction, *Simon Spero 50th Anniversary Exhibition 1964–2014: Early English Porcelain 1748–1783*.
18 Simon Spero, Spring Newsletter, April 2009, p. 2.

Chapter 13

1 Interview with the author.
2 Interview with the author.
3 Howard Ricketts in Brian O'Connell, 2013, p. 174.
4 Unpublished memoir.
5 Alastair Laing to the author.
6 Interview with the author.
7 Interview with the author.
8 Anon interview with the author.
9 Interview with the author.
10 Interview with the author.
11 Interview with the author.
12 Interview with the author.
13 Interview with the author.
14 Interview with the author.

Chapter 14

1 Interview with the author.
2 Email to the author.
3 Email to the author.
4 Wood, unpublished, n.p.
5 Wood, unpublished, n.p.
6 Email to the author.

7 Email to the author.
8 Email to the author.
9 Interview with the author.
10 John Culme email to the author.
11 Interview with Alastair Dickenson.
12 Email to the author.
13 Email to the author.
14 Martin Levy in MacLean and Hook, 2017, p. 241.
15 Mark Westgarth, 2019, p. 25.
16 Interview with Alastair Dickenson. The sale was 24 March 1982.

Chapter 15

1 Wraight, 1965, p. 25.
2 Gerald Reitlinger, 1970, p. 9.
3 Hook, 2017, p. 222.
4 Interview with the author.
5 Norman, unpublished memoir.
6 Norman in MacLean and Hook, 2017, p. 141.
7 Herrmann, 1980, p. 406.
8 Lacey, 1998, p. 124.
9 Norman, unpublished memoir.
10 *Bastian*, vol. 2, Spring 2019, p. 13.
11 Interview with the author.
12 She was appointed in 1969.
13 *The Times*, 21 December 1974.
14 Interview with the author.
15 Interview with the author.
16 Herbert, 1990, p. 220.
17 Interview with the author.
18 Interview with the author.
19 Interview with the author.
20 *The Times*, 21 December 1974.
21 Interview with the author.
22 Eskenazi, 2012, pp. 52–3.
23 Annamaria Edelstein in MacLean and Hook, 2017, p. 212.

Chapter 16

1 Thomas Grant, p. 209.
2 Norman, unpublished memoir.
3 Norman, unpublished memoir.
4 Wood, unpublished, n.p.
5 Norman, unpublished memoir.
6 Norman, unpublished memoir.
7 Norman, unpublished memoir.
8 Norman, unpublished memoir.
9 Geraldine Norman in MacLean and Hook, 2017, p. 142.
10 Herbert, 1990, p. 147.
11 Herbert, 1990, p. 198.
12 Norman, unpublished memoir.
13 Quoted in Norman, unpublished memoir.
14 Quoted in Norman, unpublished memoir.
15 Grant, 1998, p. 205.
16 *The Times*, 21 July 1976.
17 See Norman, 2014, pp. 230–1.
18 Norman, unpublished memoir.
19 Grant, 1998, p. 199.
20 Grant, 1998, p. 200.
21 Anon interview with the author.
22 Norman, 2014, p. 13.
23 Grant, 1998, p. 203.
24 Norman, unpublished memoir.
25 Norman, 2014, p. 251.
26 Norman, 2014, p. 258.
27 Norman, 2014, pp. 228–9.
28 Grant, 1998, p. 203.
29 Norman, 2014, p. 201.
30 Grant, 1998, p. 204.
31 Grant, 1998, p. 209.
32 Norman, 2014, p. 261.
33 Grant, 1998, p. 211.

Chapter 17

1 Anthony du Boulay, interview with the author.
2 Interview with the author.
3 Roy Davids and Dominic Jellinek, 2011, p. 166.
4 Interview with the author.
5 Davids and Jellinek, 2011, p. 330. The story is unverifiable but frequently told.

6 Eskenazi, 2012, p. 76.
7 Davids and Jellinek, 2011, p. 198.
8 Eskenazi, 2012, p. 48.
9 Interview for *Phillip's Magazine*.
10 Eskenazi, 2012, p. 80.
11 See Stourton, 2007, p. 277.
12 Interview with the author.
13 Davids and Jellinek, 2011, p. 16.
14 Davids and Jellinek, 2011, p. 310.
15 This version comes from Dominic Jellinek. See Davids and Jellinek, 2011, pp. 276.
16 Interview with Giuseppe Eskenazi.
17 Colin Sheaf to the author.
18 Eskenazi, 2012, p. 171.
19 Interview with the author.
20 Interview with the author.
21 Eskenazi, 2012, p. 40.
22 Eskenazi, 2012, p. 46.
23 Eskenazi, 2012, p. 60.
24 The name is a pseudonym.
25 Eskenazi, 2012, pp. 162–3.
26 Eskenazi, 2012, pp. 121–2.

Chapter 18

1 Browse, 1999, p. 155.
2 Cooper, 1977, p. 144.
3 Cooper, 1985, p. 22.
4 Interview with the author.
5 Wood, unpublished, n.p.
6 Wood, unpublished, n.p.
7 Reitlinger, 1970, p.3.
8 Hartnoll, email to the author.
9 Interview with the author.
10 David Scott sale catalogue.
11 Wood, unpublished, n.p.
12 Wood, unpublished, n.p.
13 Watson, 1997, p. 343.
14 Rupert Maas to the author.
15 Wood, unpublished, n.p.
16 Wood, unpublished, n.p.
17 Wood, unpublished, n.p.
18 Hilary Kay to the author.
19 *Financial Times*, 31 January 1973.
20 Stourton, 2007, p. 359.
21 *Bastian*, vol. 2, Spring 2019, p. 12.
22 Roy Miles, 2003, p. 105.
23 Wood, unpublished, n.p.
24 Roy Miles, 2003, p. 74.
25 Roy Miles, 2003, pp. 74–5.
26 Interview with the author.

Chapter 19

1 *Bastian*, vol. 2, Spring 2019, p. 12.
2 Cooper, 1985, p. 36.
3 Interview with the author.
4 Interview with the author.
5 Interview with the author.
6 *Bastian*, vol. 2, Spring 2019, p. 12.
7 Interview with the author.
8 Cooper, 1977, p. 139.
9 Cooper, 1977, p. 26.
10 Cooper, 1977, p. 26.
11 Speelman, 2020, p. 33.
12 Cooper, 1977, p. 176.
13 Cooper, 1977, p. 24.
14 This figure is suggested by Richard Madley.
15 Obituary Dick Kingzett *The Times*, 29 April 2010.
16 Speelman, 2020, p. 122.
17 Sewell, 2011, pp. 220–1.
18 Interview with the author.
19 Sykes, 2011, p. 100.
20 Day, 2008, p. 186.
21 Day, 2008, p. 186.
22 Speelman, 2020, p. 168.
23 Speelman, 2020, p. 168.
24 Obituary in *The Guardian*, 14 January 2016.
25 Speelman, 2020, p. 109.
26 Quoted Blanche Girouard, 2013, p. 40.
27 Information from Philip Wilson.
28 Girouard, 2013, p. 15.
29 Interview with the author.
30 Interviewed for *Leeds Oral History Project*, 2014.
31 Interview with the author.

Chapter 20

1 Interview with the author.
2 Herbert, 1990, p. 276.
3 Hook, 2009, p. 186.
4 James Miller in MacLean and Hook, 2017, p. 223.
5 Strauss, 2011, p. 229.
6 Herbert, 1990, p. 275.
7 Lacey, 1998, p. 203.
8 Lacey, 1998, p. 216.
9 Lacey, 1998, p. 219.
10 Interview with the author.
11 Henry Wyndham to the author.
12 Mason, 2004, p. 60.
13 Interview with the author.
14 Sir Hugh Roberts to the author.
15 Hook, 1993, p. 17.
16 Lacey, 1998, p. 260.
17 Herbert, 1990, p. 340
18 Mason, 2004, p. 84.
19 Hook, 2009, p. 192.
20 Strauss, 2011, p. 9.
21 Hook, 2009, pp. 199–200.
22 Interview with the author.
23 Interview with the author.

Chapter 21

1 Noël Annesley, Morgan Lecture, 25 October 2017.
2 Jennifer Tonkovich, 2017, p. 49.
3 Annesley at *Master Drawings* Dinner, 23 January 2012.
4 Day, 2008, p. 63.
5 Annesley at *Master Drawings* Dinner, 23 January 2012.
6 Interview with the author.
7 Tonkovich, 2017, p. 51.
8 Tonkovich, 2017, p. 53.
9 Tonkovich, 2017, p. 54.
10 Annesley, Morgan Lecture, 25 October 2017.
11 See Eeles, 2021, p. 384.
12 Annesley at *Master Drawings* Dinner, 23 January 2012.
13 Morgan Library Lecture, 25 October 2017.
14 Burton B. Fredericksen, 2015, p.1.
15 Burton B. Fredericksen, 2015, p.1.
16 Day, 2008, p. 297.
17 Day, 2008, p. 302.
18 *The Times*, 5 July 1984.
19 Day, 2008, p. 303.
20 Fredericksen, 2015, p. 263.
21 Fredericksen, 2015, p. 274.

Chapter 22

1 Waterfield, p. 154.
2 Shakespeare, 1999, p. 91.
3 Anon interview with the author.
4 Interview with author.
5 Email from Oliver Forge.
6 Watson, 1997, p. 62.
7 Watson, 1997, p. 60.
8 Watson, 1997, p. 116.
9 Watson, 1997, p. 142.
10 Norman in *The Times*, 22 November 1991.
11 Watson, 1997, p. 162.
12 O'Connell, 2013, p. 34.
13 Colin Renfrew, 2000, p. 11.
14 Renfrew, 2000, p. 17.
15 Lacey, 1998, p. 294.
16 Anon interview.
17 Fredericksen, 2015, p. 149.
18 Interview with the author.

Chapter 23

1 Marion True, Getty Museum internal memo, October 1987.
2 Interview with Rainer Zietz.
3 Arthur Houghton, 2022, pp. 135–6.
4 Jason Felch, 2011, p. 126.
5 Houghton, 2022, p. 129.
6 Hugh Eakin, 'Treasure Hunt. The downfall of the Getty curator Marion True', *New Yorker*, December 17, 2007.
7 Quoted Felch, 2011, p. 95.

8 Felch, 2011, p. 203.
9 Quoted Watson, 2006, p. 250.
10 Watson, 2006, p. 254.
11 Watson, 2006, p. 254.
12 Felch, 2011, p. 221.

Chapter 24

1 Lacey, 1998, p. 203.
2 Interview with the author.
3 Norman, *The Independent on Sunday*, 29 December 1991.
4 Interview with the author.
5 Interview with Zietz.
6 Houghton, 2022, p. 138.
7 Geraldine Norman and Thomas Hoving, *The Independent on Sunday*, 29 December 1991.
8 *The Independent*, 7 July 1990.

Chapter 25

1 Houghton, 2022, p. 162.
2 Houghton, 2022, p. 162.
3 Houghton, 2022, p. 162.
4 Norman, *The Independent*, 17 October 1994.
5 Houghton, 2022, p. 164.
6 Houghton, 2022, p. 164.

Chapter 26

1 Alex Wengraf, 2020, p. 142.
2 Information from the late Bill Winter.
3 Interview with the author.
4 Interview with the author.
5 Interview with the author.
6 Interview with a former director of Agnew's.
7 Interview with the author.
8 Anon interview with the author.
9 Godfrey Barker, *Telegraph*, 12 June 1995.
10 *ATG*, 2008.
11 *ATG*, 2008.
12 Corcoran owned 70 per cent of the firm, Summers 30 per cent.
13 Andrew Solomon, 1989.
14 Simon de Pury, 2016, p. 223.
15 Strauss, 2011, p. 201.
16 Strauss, 2011, p. 203.
17 Strauss, 2011, p. 205.
18 Wood, unpublished, n.p.
19 *Sunday Telegraph*, 18 March 1990.
20 Interview with the author.
21 Interview with the author.
22 Wood, unpublished, n.p.
23 Matthiessen, email to the author.

Chapter 27

1 Simon de Pury, 2016.
2 Conversation with the author.
3 Interview with the author.
4 Solomon, 1989.
5 Interview, *The Art Newspaper*, 1 January 2016.
6 Interview, *The Art Newspaper*, 1 January 2016.
7 Quoted in The Leslie Waddington Collection, sale catalogue Christie's 2016.
8 Email to the author.
9 Anon interview with the author.
10 Solomon, 1989.
11 Solomon, 1989.
12 Interview with the author.
13 Interview with Thomas Lighton.
14 Interview with the author.
15 Email to the author.
16 Michael Shnayerson, 2020, p. 68.
17 Email to the author.
18 Email to the author.
19 Email to the author.
20 Anon interview with the author.
21 *The Guardian*, 20 April 2004.
22 Anon interview.
23 Georgina Adam, 2017, p. 41.
24 Interview with the author.
25 Interview with the author.
26 Shnayerson, 2020, p. 214.
27 Shnayerson, 2020, p. 216.
28 Email to the author.

Chapter 28

1 Interview with the author.
2 Interview with the author.
3 *The Guardian*, 2 March 2008.
4 Alain Elkann, *Anthony D'Offay*, online interview, January 2017.
5 *The Guardian*, 2 March 2008.
6 Shnayerson, 2020, p. 132.
7 Email to the author.
8 Anon interview with the author.
9 Anon interview with the author.
10 Interview with the author.
11 Anon interview with the author.
12 Interview with the author.
13 Elkann, 2017.
14 Shnayerson, 2020, p.129.
15 Elkann, 2017.
16 Elkann, 2017.
17 Elkann, 2017.
18 Elkann, 2017.

Chapter 29

1 Interview with the author.
2 Interview with the author.
3 Shnayerson, 2020, p. 120.
4 Interview with Nick Serota, Ann Gallagher, 2012, p. 82.
5 Elizabeth Fullerton, 2016, p. 12.
6 He taught there 1974–88 and then 1994–2000.
7 Shnayerson, 2020, p. 187.
8 Shnayerson, 2020, p. 187.
9 Fullerton, 2016.
10 Norman Rosenthal, 1997, p. 9.
11 Interview with the author.
12 *The Guardian*, 1 August 2019.
13 Interview with the author.
14 Interview with Nick Serota, Ann Gallagher, 2012, p. 95.
15 Interview with Richard Calvocoressi.
16 Interview with the author.
17 Interview with the author.
18 Interview with the author.
19 Interview with the author.
20 Interview with the author.
21 Interview with the author.
22 Interview with the author.
23 Interview with the author.
24 Interview with the author.
25 Interview with the author.
26 Interview with the author.

Chapter 30

1 An Inquiry into the Nature and Causes of the Wealth of Nations.
2 Mason, 2004, p. 90.
3 Hook, 2009, p. 206.
4 Mason, 2004, p. 97.
5 Mason, 2004, p. 116.
6 Mason, 2004, p. 124.
7 Mason, 2004, p. 355.
8 Mason, 2004, p. 134.
9 Mason, 2004, p. 159.
10 Mason, 2004, p. 220.
11 Mason, 2004, p. 263.
12 Quoted Mason, 2004, p. 221.
13 Mason, 2004, p. 221.
14 Mason, 2004, p. 234. This was negotiated by Elizabeth Lane.
15 Hook, 2009, p. 210.
16 Mason, 2004, p. 260.
17 Mason, 2004, p. 261.

Chapter 31

1 De Pury, p. 236.
2 De Pury, p. 269.
3 Interview with the author.
4 ArtReview's 'Power 100' list.
5 Eskenazi, 2012, p. 121.
6 Adam, 2017, p. 182.
7 James (Jan) Morris, *Cities*, London, 1963, p. 237.

Image Credits

1. © Sotheby's 2. © Photograph by Snowdon / Trunk Archive
3. Photograph courtesy of Colnaghi
4. Flints Auctions 5. Photograph by Snowdon / Trunk Archive
6a. *Wall Street Journal*
6b. © R. Hamilton. All Rights Reserved, DACS 2024 7a. Courtesy of Kasmin Ltd 7b. Courtesy of Kasmin Ltd 8a. Central Press / Stringer / Getty Images 8b. Ailsa Mellon Bruce Fund / National Gallery of Art 9a. Evening Standard / Stringer / Getty Images
9b. © The Norton Simon Foundation 10. The Metropolitan Museum of Art 11. GL Archive / Alamy Stock Photo 12. Bridgeman Images 13a. © Christie's
13b. Richard Southall 14a. Lanto Synge 14b. Mallett Antiques
15a. Courtesy of *The Times* / Daily Britain 15b. © Sotheby's 16a. *South China Morning Post* / Getty Images
16b. Arts of Asia 17a. Forest Hill Society 17b. *Transcendental Meditation for Women* 18a. Pitt Rivers Museum Catalogue, 1929
18b. Bettmann / Getty Images
19. Bequeathed by Lady Sainsbury, 2014 / Sainsbury Centre
20. Royal Borough of Kensington and Chelsea Libraries 21. Courtesy of Pamela Klaber 22. ANL/ Shutterstock 23. Richard Green
24a. Stan Meagher / Stringer / Getty Images 24b. Pixel8tor / Wikimedia Commons 25. Digital image courtesy of Getty's Open Content Program 26. Antonio Pignato / Wikimedia Commons
27. Copyright unknown
28. © Hungarian National Museum Public Collection Centre
29a. Courtesy of Damian Hoare
29b. Aga Khan Museum
30a. PA Images / Alamy Stock Photo
30b. Cornell University Library / Wikimedia Commons
31. Homer Sykes / Alamy Stock Photo 32. © Christie's / Christie's London 1987 33. Niklas Halle'n / Stringer / Getty Images 34. Enter Gallery 35. © Jillian Edelstein / Camera Press 36. UPI / Alamy Stock Photo.

Index